Acknowledgements

The author wishes to express appreciation to numerous individuals who, over the last fifteen years, have helped form these concepts into practical applications and field-proven results.

This list includes, in no order of magnitude or with no impression one or the other, nor do those concerns who kindly contributed materials such as ODM, EXFO, MicroCare, and Chemtronics, agree or disagree with these applications and associated thesis.

Appreciation is extended to: Paul Blair, William Woodward, Eric Martini, Michael Hackert, Larry Johnson, Michael Schneider, Dr. Tatiana Berdinskikh, Dr. Osman Geblizlioglu, Don Stone, Mark Baranuk, Douglas Teller, Carl Walz, Gary Tyler, Dr. Gene Yon, Roger Heydinger, Mitchell Weinstein, Kenneth Putnam, Frank Giotto, James Henry, David Zika, John Lastowka, Bob Menard, George Bell, Charles Mason, Pamela Nobles, Anthony Lowe, James Henry, Kevin Wright, John Cotterill, Vincent Racine, Michael Chilicki, Pat Hanlon (RIP) , Darrell Smith, Gary Tyler, Jim Drain, Keith Hayes, Laurence Wesson, Mel Lesperance, Richard Ednay, Kim Teesdale, Bill Johnston, Kirk Donley, Dan Morris, Brian Teague, WP Beverly, Paul Looney, Michael Yeilding, Dan Morris, Earle Olson, *my wife, Lanet, and thousands of technicians, dozens of field sales engineers and manufacturer's representatives who have input their applications and experiences though many contacts and practical associations. You know who you are!*

Likewise, appreciation is extended to AT&T, BICSI, CenturyLink, Comcast, FTTH Counsel, IMSA, iNEMI, OFC/NFOEC, OSP-EXPO, SAE-Aerospace, SCTE, and Verizon for speaking and technical platforms to interact with industry professionals on all levels. *Thank you, all.*

What you can expect!

✓ **A better understanding of
"Why and How to Clean & Inspect"**

✓ **What works ... What Does Not and "Why?"**

✓ **How to improve what you are using now.**
- **Cleaning technique**
- **Inspection techniques**
- **What happens when you do not have
inspection (It's not what you may think it is!)**

✓ **A "Call for Best Practice"**
- **Existing standards: are they not?   Why?**

I.) What Are We Cleaning?

- **New Information on connections**
- **Understanding the "science of debris and contamination"**
- **What is the 'role' of a standard**

II.) How do 'we' know it's clean?

- **What is an OTDR?**
- **What is a Light Source?**
- **What is a Fiber Identifier?**
- **What is Video Inspection?**
- **When and how to use each instrument**

III.) How to clean it?

- **What cleaning tools are available?**
- **Should I expect they all work the same way?**
- **What works, what does not,**
- **What is Best Practice?**

Often separated into segments of "OEM versus OSP", "commercial services versus aerospace", or "Telco versus CATV" existing fiber optic standards tend to be portrayed as maximum requirements when, in the face in ever increasing speed and capacity, these are actually "minimum requirements". This is not acceptable for *any* rapidly evolving science. To develop and commercialize a total view the Industry requires research and development that considers a new standard type that evolves continually and not bound by a structure of updates every 5 or 10 years interspaced by lengthy composition processes.

By and large unchanged since inception, the tenets of IEC 61300-3-35 have influenced and developed standards that includes: a.) the importance of cleaning the fiber optic connection, b.) the concept of diameter of debris or contamination, c.) the area of the end face to be cleaned, and, d.) a series of procedures and implied products. This may not be adequate or accurate.

There are myriad types of debris and contamination. Some are "dry" and others are "fluidic". Often times the debris on a connection is present in "combinations". Regardless of type, the cleaning procedure should strive to be a first time process and not as currently considered: multiples of procedures therein when one procedure and then another is an attempt to return the end face to an acceptable condition.

Dry, fluidic, or combinations of two (or more) contaminations have height. Some types of these are noted in existing standards but, by and large "easy to remove" soils define standards when "worst case" should lead to "best practice". A commonly used cleaning technique can create fluidic contamination emanating from the area outside the view of most-used video inspection and software analysis. Another technique can create a static field that attracts additional debris. Still others are complex soils, or unidentifiable and unknown. Not considered in existing standards: *all contamination has height and connectors themselves are three-dimensional structures.*

As well not fully considered is that an ineffective cleaning procedure may result in *removable* debris mis-characterized as an *un-removable* artifact. There is also a sense IEC 61300-3-35 standard is relative to both production line applications (OEM) and field service (OSP) applications. These are two separate worlds and while the OEM should have interest in how a product is deployed...this is not always true as cleaning instructions are vague.

Existing standards for end face inspection limit the field of view to zones within a two-dimensional field-of-view limited to (video) magnification ranging between 100x-400x usually measured at 250-300mu radius of the core. By these standards, the end face is divided into three or four "zones", all contained on the horizontal axis.

Of course, a connection is a three dimensional structure with a perspective beyond this limited field of view and extending through the complete 'geometry' of the connection. (FIGURE-1)

However, there is an area outside this limited field-of-view that includes the remainder of the "horizontal axis ferrule" and intersects with a "vertical axis ferrule" to create the complete three dimensional assembly. Surrounding the ferrule are various plastic, metal or ceramic assemblies that comprise the connector.

Certain types of debris outside the field of view (Zone-3) can migrate into the core. This type of migration would most likely be a fluid contamination. 'Fluidic' contamination can be excess cleaning solvent, an oily type contaminate, or even condensation contamination wherein a connection is moved from one ambient to another.

"Dry" contamination comes from various ambient. (TYPE 1 and TYPE 2). Fluidic contamination is depicted in image TYPE-3. Some are also unidentifiable (TYPE-4). There are often-used cleaning techniques that wipe or spray-apply excess cleaner in the initial phase of cleaning. Excessive cleaning solvent can be harbored in Zone-5 and migrate in the post cleaning, post-inspection process. While video inspection at 100-200x creates a wider field of view, thereby enabling inspection outside "Zone-3", most video inspection instruments are not capable of discrete image resolution sufficient to identify the debris.

While it may be impractical to change a standard such as IEC 63100-3-35 in the short term, it is practical to heighten awareness and establish a best practice standard that exceeds this baseline. In fact, for rapidly evolving technology "standards" may best be separated into: a.) "minimum baseline" and b.) "technology advancing" as (in the instance of fiber optics) transmission equipment and deployments outpace the time between review and publication. As well, "internal standards" also can be established.

Since the fiber optic ferrule assembly and the debris are three-dimensional, this science becomes important for all fiber optic precision cleaning applications. A jumper-side ferrule inserted through an alignment sleeve can both transfer debris from one end face to the other as well as capture and transfer debris on the alignment sleeve. Certain residues from the adapter itself can cross contaminate connector components. These phenomenon are possible for both direct contact and expanded beam.

Interaction of contamination and debris and the geometry of the connection are essential awareness. As capacity and speed increase, implementation of effective cleaning techniques that do not cause recontamination are critical to future connectivity. Inspection is mandatory. Cleaning procedures that work more reliably, not only based on convenience, are essential to best practice.

In search of:
Best Practice

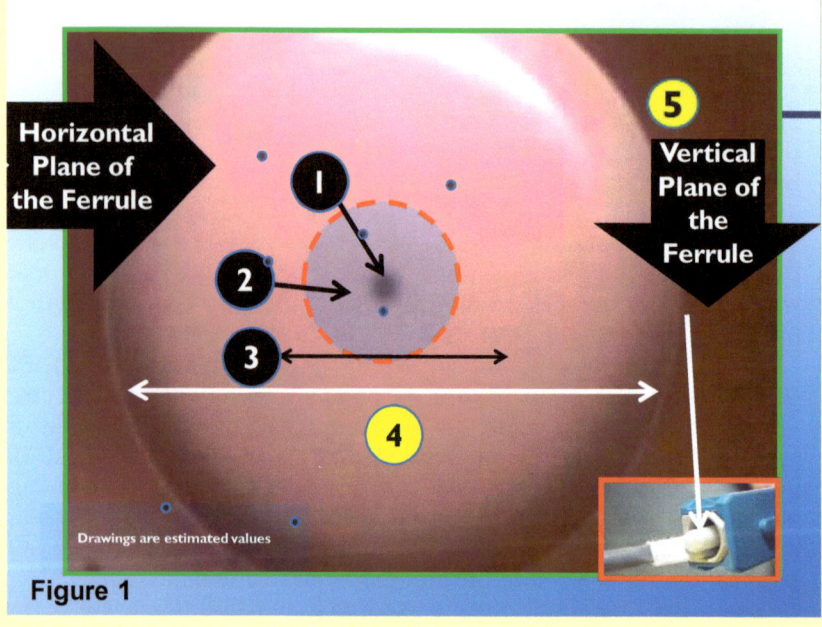

Horizontal Plane of the Ferrule

⑤ Vertical Plane of the Ferrule

Drawings are estimated values

Figure 1

What is the best way to properly clean the widest range of possible contamination from this three dimensional structure?

OSP technicians, realizing that they may not have the most effective means of cleaning, either downplay the task or work tirelessly to precision clean when their analytical skills to interpret, an OTDR trace (for example) would be a more meaningful use of time. I believe a cleaning standard is based on a process that works from "worst case" and leads to best practice, rather than 'easy-to-remove' contamination. Current practices lead to design and production of tools that are convenient, removing simple soils and not necessarily "best practice".

In addition to the range of dry debris types, Static Field Contamination is also possible. This is termed *ESA* (Electrostatic Attraction) by the ESD Association Standard S-2020. ESA is likely to occur when the ambient temperature ranges in the 20-50F range or >95 with relative humidity in the 35-60% range. Likewise, ambient in the >90F range can promote ESA.

Static field contamination should be considered as one of many debris types. Generally speaking, there are four ways to control static field contamination: 1.) is use of an air ionizer, 2.) use of a grounding strap, 3.) use of static topical coatings, 4.) dissipation by introduction of a precision cleaning solvent.

The images with common dust and an uncommon eyelash, (Type 1 & 2) were captured at 72F and 50% relative humidity. The actual static field was only 78v and attracted the light dust to the end face. At <38v there was no attraction of debris.

In the world of production ESD, typically there is a 2000v/in limitation. Therefore, ESA of debris at values <100v is a significant number. These image depict examples of various types of "dry" debris, "fluidic" contamination, combinations and the unexpected!

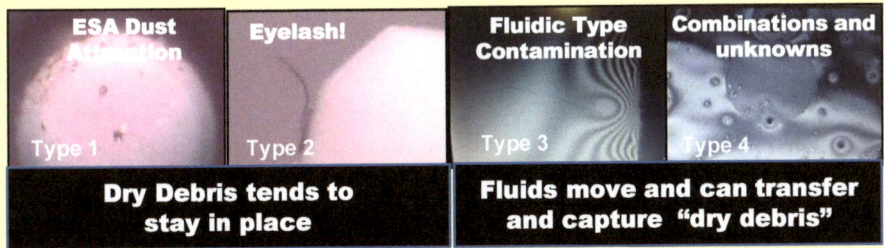

Is two-dimensions Best Practice?

Why are you reading this?

- If You are a Designer…

    • If You are an Installer…

- If You are a Service Seller…

    • If You are a Trainer…

- If You Happen to be an OEM producer of equipment…

There are fiber optic deployments in your future!

➢ **Category Cable Connected to Fiber Optics**
➢ **Wireless systems integrated to Fiber Optics**
➢ **Discrete FTTx**

You may also be confused by claims and counterclaims regarding individual products. Some of them work and others don't. This book is written to clear the air about these things!

There are many pressures including sales budgets! Some training programs are based on articles that tend to restate existing standards and don't 'look outside that box'. There are products based on these standards with little will to change production runs. Some written work is in a publication because an individual or corporation paid to have it there! Some standards groups have hefty fees associated with participation while others encourage all contributors and (understandably) may limit voting rights.

This book and my research is 'vendor neutral' and is on going. Yes, there is a fee for the book, individual research and training projects. I try to participate on Internet Forums …and this is, of course, free to all *(as much as it may be a free-for-all!!)* Please communicate your specific needs: I'll always do my best to respond and always return your inquiry with an unbiased response.

1st: <u>What</u> (connector type) <u>am I cleaning</u>?

2nd: <u>What & Where</u> is the contamination?

3rd: What process, tool, or device will remove (ANY AND ALL) contamination?

When we think about it, these are the three questions asked every day by tens of millions of individuals who make decisions about "cleaning". These are the same questions raised for the last 3,000 years beginnings with The Babylonians who are credited with inventing "soap" from sheep tallow, to the recent elimination of CFC's throughout the 1990's, to the current generation of "aqueous" cleaners that replaced hydrocarbon cleaners invented by the Chinese in the dawning years of The First Century.

For fiber optics connections the question is about the finest form of cleaning. I term this "precision cleaning". It is inspired by procedures used in high-value clean rooms used to produce "solid state' components and sterile environments: implanted medical devices. Of course, for fiber optic applications the "environment" can be a relatively clean data center, or a dusty room in an apartment complex where Fiber to the Home is being introduced. A fiber optic system may be in a military theater, or 20 feet high strung between poles or a right of way through a pasture, along a river basin, trans-oceanic or trans-continental. The venues are infinite.

Cleaning a fiber optic connection is the critical first step to assure one thing only: the end user is delighted with the performance of the magnificent medium of fiber optic transmissions. There is nothing faster than the speed of light. The miracle that is a fiber optic transmission passes along a slit of glass barely greater in size than a human hair. Many of us understand how a dirty lens on eyeglasses "refracts" our vision. Few are not annoyed by a windshield that has contamination on the outside...and residues on the inside. Windows on an office building distract our view. These are huge surfaces what are clearly visible. In many instances a fiber optic connection is not seen before or after it is cleaned.

All of these considerations, and far more, are presented to you on the following pages.

What is Best Practice?

As wide a range of contaminants as imaginable should be catalogued. One such evaluation was made in 2006 with a series of contaminates from Cisco®. The "Cisco Series" ranged from Arizona Road Dust to dried water. Graphite, Simethicone and Dryer lint were included. In recent times, testing was suggested on dust from Afghanistan, which is remarkably finer than ARD. Beach sand is coarser than ARD: any and all factors are relevant to effective cleaning and inspection techniques. "Beach sand" may be relative to storm damage, dust from feed lots, salt water and myriad others influence network, data center, and all FTTx services.

A second series of laboratory evaluations was conducted in 2010 at my ITW Chemtronics® Laboratory. In this study (recorded on video) a proprietary HFE-7100/IPA formulation and a second proprietary HFE-7100/non-IPA formulation were contrasted against a a proprietary precision hydrocarbon. compared two different HFE-7100 formulations and a proprietary precision hydrocarbon removing a wide range of complex contaminants. One cleaner, the precision hydrocarbons, clearly outperformed the others. A copy of this study is available for your interest: it is a large video file!

The debris included Arizona Test Dust as representative of a dry-type and vegetable oil as representative of a fluidic-type contaminant. In all there were a 10 variety of ten neat contaminants. (neat is defined as pure, full-strength, unadulterated). The dry contaminants and fluidic contaminants were then mixed into combinations of types. The intent of the study was to create worst case debris that would lead to best practice for OEM (production) and OSP (installation) as well as OEM that may have field service or installation capability.

Combinations of debris such as human body oil and Arizona Test Dust, mineral oil and carbon black from field service at coal fired power plants such as TVA. Silicone oils, dust other than ATD such as found in other desert regions, and a host of diverse types of debris should be considered when creating a sense of future proof precision cleaning against new transmission technologies. A localized, self-standardization practice is suggested for all who work with fiber.

A third extensive test series was conducted in December-2014 that compared and contrasted current cleaning tools through a wide range of contamination to baseline "best practice".

IEC 61300-3-35 and all derivative standards are based on removal of relatively simple debris. Therefore, the fundamental understanding of precision cleaning a fiber optic connection tends to be overly simplified. This over-simplification leads to confusion for both field service technicians as well as OEM personnel all seeking to increase output and reduce costs.

What do we mean when we say "Best Practice"? *

"Best practice is a method or technique that has consistently shown results superior to those achieved with other means, and that is used as a benchmark."

A "best practice" *can evolve as improvements are discovered*.

Best practice is considered a means to describe the process of developing and following a standard way of doing things *that multiple organizations can use*.

There is nothing more important than a road map and nothing worse than one that is outdated!

Existing standards for fiber optic inspection and cleaning were first penned in 1998 and updated in 2008. It appears the next update will be around 2018.

There have been significant changes in fiber optics since 1998 and 2008…and these will continue past 2018. While these standards are written by some of the finest minds in our Industry…

Unfortunately, many 'perceptions' are not in line with accepted tenets of precision cleaning. Important aspects are missing and these include the infinite types of debris that may be present in OSP. Existing standards are conceived for controlled environments as a production floor which may have filtered air and far less complex types of debris. OSP operations demand a high level of precision cleaning and this will continue as deployments enter "ultra-speed and capacity" at multiple Gb/s entering Tb/s and beyond. This is the time to "future proof".

Let's discuss existing standards and how they can be developed from their "minimum requirement" to "best practice"

* Edited from:
1. Wikipedia® Search 24 July, 2015
2. www.whatis.com.
3. Cambridge English Dictionary
4. www.infoentrepreneurs.com

A fiber optic transmission is not unlike looking through a window.

That "window" is called an "end face". It's present on all fiber optic connectors in various configurations.

A fiber optic connection must be "beyond clean" … … it must be "precision cleaned".

For regular cleaning functions we're able to see what we're cleaning…very often for fiber optics we *either don't "see" what we're cleaning or we do not see 'enough' to know if it is actually clean.*

"Precision Cleaning" is the type of work done in semi-conductor production, food processing, or medical device manufacturing. Those are a few clean-room applications, all, in strictly controlled environments.

For fiber optics, we must be as clean: in a storm at the top of a bucket, in a vault, hot & dusty storage closet or railroad right-of-way or combat theater … some place this morning and another this afternoon!

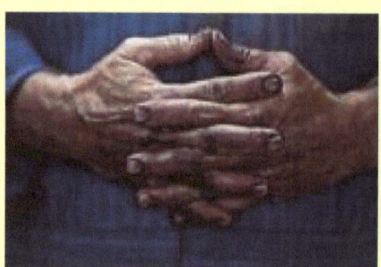 

Cleaning greasy hands may be different than cleaning a spill on a tile floor … and… *precision cleaning a fiber optic connection is also an <u>applications specific decision</u>!*

Deployments can be a harsh, diverse environment with myriad debris and contamination types…anywhere & everywhere.

Single mode…Multi Mode…Category Cable or Coax…all are being pushed to higher transmission speed and capacity limits. Only a fiber optic connection negatively may be *influenced by a contaminated surface*.

Actual test jumpers returned to the factory… as "defective; warranty return".!

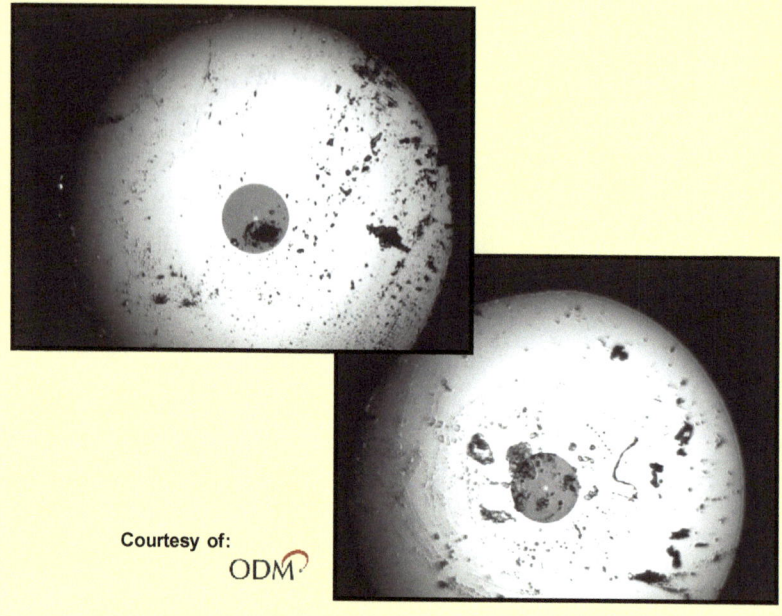

Courtesy of:
ODM

"My fibers were giving me 'fuzzy' readings…"

The question of cleaning is often confounded by the reality that many simply do not understand not only the need to clean the surface, but also what is actually clean…and what is not!

Manufacturers regularly report these incidents: high-value circuit boards are returned for analysis or "warranty repair" that only need to be cleaned. I once saw a half-full 55 gallon drum of "bad jumpers": the center manager agreed to have me clean them and 9 out of 10 were easily returned to service. These were 100' long runs. *How does this happen*?

Why is this topic so difficult for some and straight-forward for others? One reason is there are many points of view and products marketed and promoted. Existing standards are themselves promoted as "maximum values" as 'auto-detect' software is actually a "minimum requirement'. There is a conflict between science and commercialism that can and often does cloud reality.

All of these topics unfold on the following pages.

i. It all seems very logical

ii. *Aren't we doing this now*

iii. Why are we not applying proven tenets to our science

About 5,000 years ago The Babylonians invented the first substance we know as soap! As early as 345 C.E. the Chinese used oil and in the 800-900's the Babylonians paved streets with tar. It's thought one or both civilizations distilled the oil into a solvent for cleaning or illumination. (Wikipedia)

In the early 1800's near Titusville, Pa. Edwin Drake drilled the first oil well in the USA. In this decade we have seen oil shortages turn to oil surplus as new technology evolved. These advances are paralleled in other rapidly evolving technologies and this includes fiber optic deployments on all levels. Hydrocarbons are an important resource for many products including precision cleaners. There are many oil producing areas on the planet and numerous techniques to acquire this valuable resource. Hydrocarbons and their derivatives are one type of solvent necessary to clean a wide range of contamination from a wide range of surfaces.

In 1988-89 Industrialized nations created International Clean Air and Safety Standards that resulted in elimination of "cornerstone" chemicals that industrialized the planet for more than 100 years. This important international action lead to other series of cleaners that have been in the market for more than thirty years. Some of these are now being replaced. Among the most contemporary are "aqueous cleaners" which are used for electronics, electrical applications and other industrial procedures...as well as fiber optics. Aqueous cleaning may become 'the future of fiber optics'.

For over five millennia various surfactants, chemical solvents and associated wiping materials have been formulated, updated, and (even) discarded one in favor of the other.

There is immense science to the concept of cleaning. New and advanced products are regularly introduced for various operations. Some industries have formal standards while other products are deemed "successful" by their tenure in the market!

This book discusses cleaning and inspection products and procedures seeking a "best practice" to future proof designs and installations now...and into the future. A different standardization process is recommended and outlined.

i.   It all seems very logical

ii. Aren't we doing this now

iii.   Why are we not applying proven tenets to our science

In the optical fiber industry we do clean, attempt to clean, and, oftentimes ignore the lessons of history and science !

A few weeks ago I was invited to one of the commercial training seminars that rotate through our landscape. Every topic associated with fiber optics was on the syllabus …
<u>except cleaning fiber optic connections.</u>

It's not "too soon" to bring some clarity...and a sense of urgency to this discussion!

"Hey, Bob...do you know how to clean a fiber optic connector?"

"Sure, Bill...most times I clean it under the collar of my shirt …
<u>that's cleaner and better than the sleeve!"</u>

This is the short version of a real discussion!
The name and location of the event are with-held to protect reputations!

The reality is there are so many ideas that 'reality' can often be obscured by marketing programs, sales goals and just about everything *other* than "science". Just as there is immense science to cleaning products such as Dawn® dishwashing soap (generally considered the best in class) or Tide® for clothing or Simonize® for automobiles.

The fiber optic industry has yet to reach a ground level and identify a "best practice". Why?

Our technology has evolved so quickly, and continues to do so in such a way that once we find the 'ground floor' there is a new ceiling!

i.   It all seems very logical

ii. Aren't we doing this now

iii. Why are we not applying proven tenets to our science

This is the time to establish "Best Practice"

and

Eliminate inspection and cleaning "mythology"

Why?

As a designer, contractor, trainer...you must be prepared for the future. This includes properly inspecting and precision cleaning a fiber optic connection.

- If you are a network designer or member of a specification committee, you have an important and oftentimes understated role. As a network designer your immense invaluable talent and knowledge is traditionally placed on a drawing board and transferred to a blue print. There is one critical factor not included: *applications specific instructions on how to properly clean and inspect the connection. I urge designers to specify an "applications specific" cleaning process for all your work.*

- Existing standards are written and contemporary automatic detecting software programs test equipment for a limited range of contamination. With deployments of fiber optics expanding into all geographic regions and many new business types...these standards *became* maximum requirements for deployments that are ever changing and expanding.

- The chances are very-good a contemporary fiber optic connection is not being adequately inspected of cleaned for the capacity it is transmitting. The chances are 'even-better' that components (jumpers and high-value circuit cards) are being replaced when adequate cleaning would result in time saved. Yes...time is $$$.

- As a contractor, your bid is your income. Knowledge of the best inspection and cleaning procedures assures your reputation.

- As a trainer, updating to the most current information is critical not only to the class coming in this week, but also for those students in the past. Updating and continuing education means your former students know all they need to be successful.

- As a member of a specifications committee, knowing the difference between "science" and "sales" can be difficult. You have many responsibilities and keeping current is always a challenge.

This book is written for all of you...including OEM manufacturers who have an essential responsibility to assure their products and technical manuals are up to date to support old and new customers.

What Does It Mean?

"Future-Proof"

1. The <u>process</u> that seeks to anticipate future developments.
2. Actions taken to minimize negative consequences.
3. Seize opportunities
4. Change old perceptions into new realities

It was not so long ago that telecommunications transmissions were measured in baud rate and 56K baud was pretty fast stuff...

- *In 2010, Nippon Telegraph and Telephone Corporation, Fujikura Ltd., Hokkaido University, and Technical University of Denmark demonstrated ultra-large capacity transmission of 1 petabit (1000 terabit) per second over a 52.4 km length of a 12-core fiber.*

- In July-2014 Technical University of Denmark set a new record of 243Tb/sec <u>over one fiber.</u>

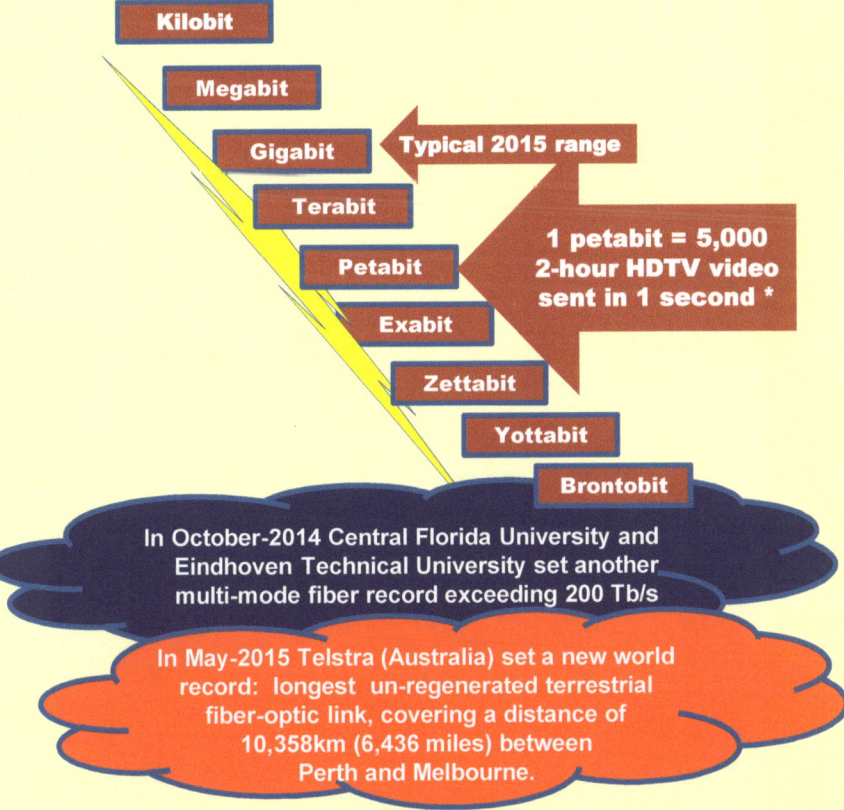

Kilobit
Megabit
Gigabit
Terabit
Petabit
Exabit
Zettabit
Yottabit
Brontobit

Typical 2015 range

1 petabit = 5,000 2-hour HDTV video sent in 1 second *

In October-2014 Central Florida University and Eindhoven Technical University set another multi-mode fiber record exceeding 200 Tb/s

In May-2015 Telstra (Australia) set a new world record: longest un-regenerated terrestrial fiber-optic link, covering a distance of 10,358km (6,436 miles) between Perth and Melbourne.

An improperly cleaned connection contributes to insertion loss, time loss in test, inaccurate testing.

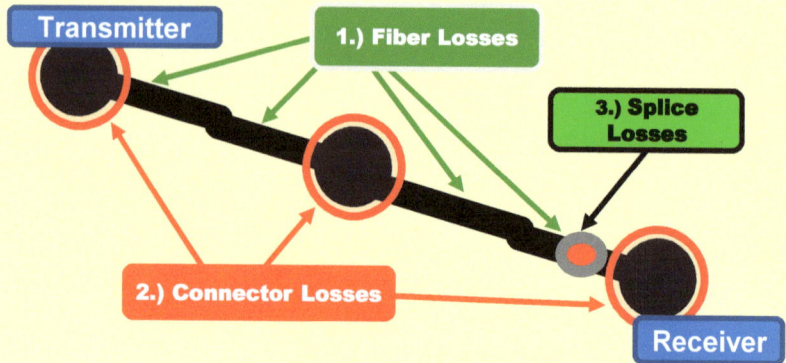

Industry expert Jim Hayes, of the Fiber Optic Association (FOA) defines: "insertion loss is the direct loss of a (signal) along a fiber optic cable" through a measurable segment.

Insertion loss is a "design" sum total of three components: 1.) Loss of light as it passes along the length of the cable: this is noted as a Fiber Loss, 2.) loss at a connector: this may be many thousands and an aspect of connector design as well as contaminated or improperly cleaned surfaces, and 3.) there are losses as the cable is spliced. The calculation is measured by test equipment in "db" and essential to every fiber optic installation.

A properly cleaned connection is critical to a low insertion loss budget. Knowing it's done properly is called "inspection"

This simplified diagram depicts a "transmitter" and a "receiver". At these points (2.) there are hundreds to thousands of fiber optic connections, each with two end face and surrounding areas terms as their "geometry".

A third spot is the point where there is a splice. Old timers speak of splicing with acetylene torches in a painstaking task that could take "all day" for one splice! Contemporary fusion splicers perform single and mass splices on ribbon fiber in a few seconds at increasingly low losses.

There are hundreds, if not thousands of fiber optic connections from "transmitter to receiver"...many splices!

At some point any one can become a "problem child"!!!

The work you are doing in 2015 and into the foreseeable future is:

1.) Impacted by necessity: PRECISION CLEANING AND INSPECTION OF THE FIBER OPTIC END FACE

2.) <u>How</u> the cleaning task is actually performed.

3.) The need to <u>properly</u> inspect the connection. What happens when inspection is not possible?

THIS IS THE TIME TO LEARN, TRAIN AND RETRAIN:

- Fiber optics as a stand alone

- Fiber optics interconnecting with Category Cable

- …as all interconnect with wireless operations.

Why?   Of all functions your personal training is critical to your success. Likely you have invested time and money to obtain certifications and other credentials. These skills require practice and practical field applications.

However, once learned your skills require updating as not only a craftsman/installer, or designer, but also but also as a trainer. Continuing education is essential.

With rapidly-evolving technology such as fiber optics and advances in Category Cable, something as small as cleaning and inspection can impact your day-to-day work…and re-hire for more.   Yes, this is that important!

It's not clear "why" proper cleaning and inspection has not taken a "front row seat" in the fiber optic industry! My theory is that until this time…precision cleaning did not influence low capacity transmissions as it will now and well into the future. Verizon's deployment of FiOS® will be recorded as a 'starting point' to the 'need to clean' as video was added to voice and data in their FTTp deployment.

With once theoretical ultra-speed and ultra-capacity now becoming reality, more than ever, 'the weakest link in a fiber optic transmission is the condition of the end face at not only time of test, but also time of actual transmission which may be at *some time* after test.   Doing it right is very meaningful to the future of the Industry as a whole.

ELIMINATE CLEANING MYTHOLOGY.

The three aspects of cleaning include: 1.) knowledge of what you are cleaning, 2.) the type of contamination present, and 3.) the ways and means to remove it. Each of these are a science unto their own. While this seems simplistic, the reality is that existing standards are based on IEC 61300-3-35. This important work is fostered by equipment producers and is directed to the production line. While the OEM does have a strong interest in what happens to the equipment once it is produced, there are few who have been willing to change the document from its existing form and format. Even installation instructions are vague in regard to cleaning and inspection.

The strengths of IEC 61300-3-35, and associated IEC TR-62627, TIA 455-250, SAE AIR 6023 and Telcordia GR2923-Core are detailed in this book. All intertwine with each other and there is little to no new ground in each writing.

To create a new baseline it's essential to understand a little chemistry, a little geology, a little meteorology as well as sciences of test and measurement. My goal is to accomplish this in a way that doesn't cause you to take a short nap! The understanding of these interactions is critical to the end goal of "best practice".

These include the sciences behind contamination, inspection and precision cleaning. Also, it's important to have a basic understanding of fiber optic test equipment. Which device does what? Do they all provide the information you need and is there more than one that will clearly tell you if the connection is actually clean?

Doubtless you have a favorite cleaning tool. How does it really work? Is one "product" better than another? What are the differences? Can one be made to perform better?

As is the case with our "buddies, Bob and Bill" there is much word of mouth and "we've always done it this way". The Fiber Optic Industry suffers from much misconception and misperception. Yes, it is a technical science that requires specialized training and expensive equipment. However, the equipment will not test properly if the connections are not cleaned properly. There are many instances where simply cleaning the connection results in "success".

The "problem" is that there are so many directions and misconceptions it has become "impossible" to outline one procedure. There are strong commercial interests.

In future times, historians will cite Verizon's experiment with FiOS® as a defining moment. Until 2004 or so, the buzzword was "broadband" and speeds and capacities were measured in kilobits per second and baud rates! I recall banners promoting 40 megs/sec in 2005 at a SuperCom® (an important trade event that ran between 1988 and 2010.). In a very short time the industry has transitioned from what was once "theoretical" to "practical". Even an event, as SuperCom, highly anticipated and widely attended, lost momentum.

What has not kept pace are cleaning and inspection perceptions! For a time speed and capacity had not demanded 'pristine clean'. *Cleaning on the inside of a shirt collar was as good as any! Verizon's deployment of video changed these perceptions: something as simple as a soiled connection meant hours of work. This was resolved when ITW Chemtronics® implemented a new cleaning procedure.* I am fortunate to have invented the concepts: there were many who influenced this with significant inputs and contributions making it truly an industry-wide effort.

Human nature and change are not easy things to accept. Standards do not define a "best practice" but rather are satisfied with minimum requirements.

There are many commercial interests, claims, counter claims...and product choices. *"What if" ... any product could work better with a "process change"? "What if" discussions about convenience and transportation of goods took a back seat to what worked best?*

In 2006 Cisco® published a study that expanded the IEC standards of "simple soils" to more complex ones that included difficult to remove debris such a metal shards, graphite and dried salt water. Between that time and 2014 no further study was completed until a work I performed that included not only "dry debris", but also "fluidic contamination" and the two in "combinations". The work went on to prove that most popular cleaning products have better performance when used is one specific process.

"If the end face is not clean, or properly cleaned ... the light will be refracted, reflected and lost as it passes along the fiber optic cable." [a]

(a) Bill Woodward: *Fiber Optic Installer Certification Guide*
Sybex-2015

The need to precision clean a fiber optic connection based on science: not convenience, ease of transportation, low cost, high cost or other commercial consideration than preservation and expansion of the fiber optic industry and your individual interests.

There are competing factors: one of them is the amazing advances of high speed and capacity over category cable. The other are the advances and acceptance of "wireless" transmissions. There are not winners or losers. Each has an important and critical role as 100 year old copper networks and installations are updated or replaced.

There "problems "still to solve" regarding fiber optic transmissions:

1.) Technology moves faster than evolution of standards of inspection and cleaning.

2.) Inspection and precision cleaning procedures were not 'base-lined' adequately

3.) We have trained and possibly 'under-trained' tens of thousands of technicians

"Inspecting and precision cleaning this surface ... no matter the configuration ... is the essence of *supporting the current and foreseeable technology requirements for at least the next 30 years"* [b]

The condition of the "end face" not only at the time of a fiber optic transmission, but also, at the time of post-cleaning and post-inspection is of critical concern in our search for "best practice".

After a training meeting, 'on more than one occasion' a technician would report "I cleaned it, but came back and it looked like I was a lair!". *How can this be possible*?

As is the case with most things, there is not likely to be one response and that's a good thing in our search for best practice! Remember, there are three aspects to cleaning (anything)...1.) what am I cleaning, 2.) what is the contamination, 3.) what ways and means are available to perform this job?

b.) John Cotterill: JSC Aeroptics. Somerset, UK. April-2015

What am I cleaning?

There are many connector types. There are also "transceivers" [3] which are the connection points for a fiber jumper to the transmission electronics.

In this discussion we are not going to discuss which is better than the other. Any and all transmit a fiber signal as light. Any and all are influenced by contamination. Any and all must be properly cleaned to maximize performance and obtain lowest insertion loss calculations. Any and all have an "end face". [4] . Some have pronounced "geometry" with deep recesses and ferrules. [5] Others have more "shallow" recesses. [6] Some of these areas are considered and termed as "zones" by international standards. Other areas are not defined by international standards.

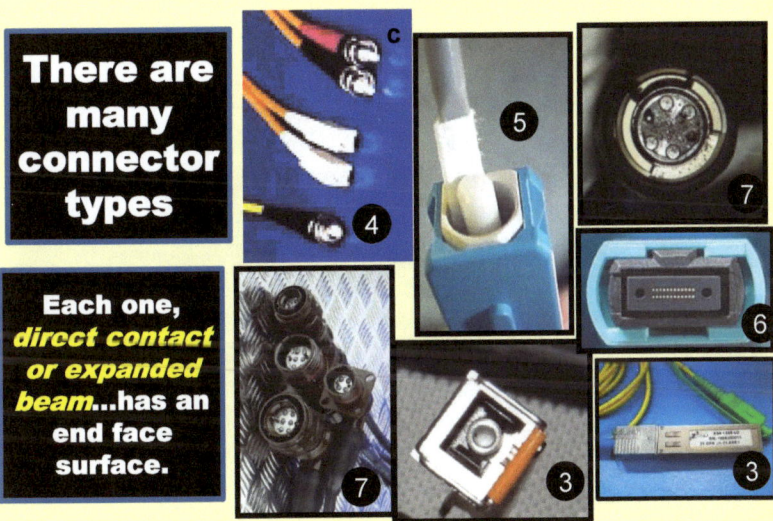

The 'end face' is the 'business end' of the fiber cable. If it is damaged or contaminated there is inevitable insertion loss.

Above there are two types: "direct contact" and "expanded beam". Direct contact types are designed as the term implies: there is a tactile connection between two "end face", or, multiples as seen in [6]. The "expanded beam" [7] has a lens over the actual fiber. Since light is transmitted between two lenses, there is no physical contact. [c.]

Existing cleaning and inspection standards consider a fiber optic connection only in terms of an end face as a two dimensional structure. Over the last thirty years all training and (cleaning/inspection) product development is based on this fundamental misperception. We exist in three dimensions!

c.) **White Paper: Expanded Beam & Physical Contact Fiber Optic Connectors.**
Edward Siminoni/James Douthit: Amphenol

The IEC 61300-3-35 Standard
View of an End Face

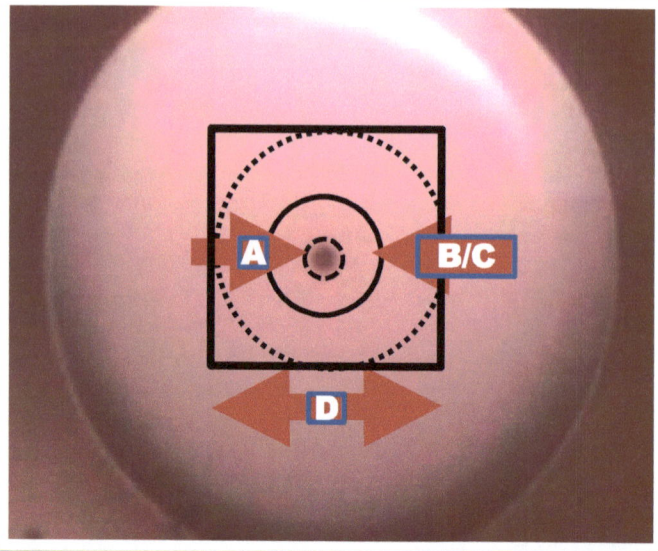

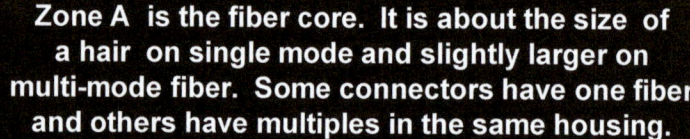

Zone A is the fiber core. It is about the size of a hair on single mode and slightly larger on multi-mode fiber. Some connectors have one fiber and others have multiples in the same housing.

Zone B/C is the 'cladding area' and 'epoxy ring'. It is defined on the inner diameter as Zone-B and Zone-C on the outer diameter. It's about a 125 micron radius from the 'core'.

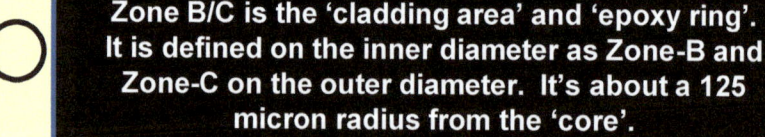

Zone-D is the visual limit of most video inspection. It's often termed the contact zone". 400x video inspection (often required to discern debris types) limits field-of-view to approximately a 250-300 micron radius of the 'core'. (on a 2.5mm connection)

Isn't this "good enough"?

Cleaning (anything) is an interaction of the surface to be cleaned, type of contamination and means to remove soils.

The IEC standard view ignores a critical aspect of physical science: *we exist in three dimensions*.

Author's Note: This image is hand drawn.
The perspective is a close approximation.

26

The "weakest link" in the magnificent medium of a fiber optic transmission is the condition of the end face at the time of test and transmission

Fiber Optic Transmission Technology moved faster than evolution of standards for inspection and cleaning.

1.) Standards became "stranded in time" as they only are updated every ten years.

2.) When existing standards were written, Inspection and Precision cleaning procedures were not 'base-lined' adequately: video scopes were once "bench top" scientific microscopes and Clean Air Legislation drove the Industry to use 99.9% IPA.
We've come a long---long way.

3.) "Re-train to FutureProof" tens of thousands of technicians….

In the "Search for Best Practice" for a rapidly evolving technology the old way of standard creation may no longer be relative. This is troublesome as there are products and business interests based on these works. However, at some point in time either the standards are changed or, self-standardization occurs. I propose a new standard type that evolves with Industry advances, created as a "blog" on the Internet, vetted by Industry experts with input from those who might not be able to attend a formal standard meeting. Best Practice advances best-interests of the Industry.

The product development landscape and its history is littered with false starts: BetaMax® and VHS®, reel-to-reel, 8-Track tapes and cassettes are only a few "sure things". What do you remember? We have to se sure fiber optics does not become one of these "memories".

How do we know if a connection is "clean"

By Standards

- **IEC 61300-3-35**
- **TIA 455-240**
- **Telcordia GR 2923-CORE**
- **SAE AIR-6021**

Strengths:

1.) Traditional baseline

2.) Composed by the brightest minds in the Industry on a global scale

Weaknesses:

1.) Ten year cycle

2.) Limited criteria

3.) Based on production cycles not technology evolution or OSP

4.) Committee sessions tend to be closed-door.

By Testing

- **Fiber Identifier**
- **OTDR**
- **Power Meter**
- **Visual Fault Locator**
- **Direct or Video Inspection**

Weaknesses:

1.) Which instrument do I need?

2.) What works best for what?

3.) Confusion based on commercialism!

Strengths:

1.) High quality results from each instrument

2.) Costs have moderated to make more accessible.

3.) Rental Programs

This may not be good enough!

How do we know if a connection is "clean by standards"

HERE IS ANOTHER CONCERN:

Existing standards are studied and published for the production line and transposed to field service applications. The production line is a controlled environment.

Field service environs are infinite. Those tasks are performed by both highly skilled and casual craftspersons who may, or may not be given the required skill sets to identify 'applications specific' inspection or cleaning problems. These "problems" may relate to a "here and now" physical environment of 'field services' compared to one that is otherwise climate-controlled, or highly applications specific by repetitive motions common to a production line or specific production run.

In many ways there should be two standards: one for the production line and one for field services. Currently, (and as planned for the future*), IEC standards, which can create updates for other groups such as TIA, Telcordia, SAE, ARINC and numerous individualized benchmarks, will carry over to field services. This existing 'one for all' is not well considered. In many ways production line work *does* become a field service application as equipment is deployed. However, rarely are actual cleaning procedures for equipment installation actually detailed by the manufacturer likely because of the incalculable sites of the actual deployment.

Cleaning methods and procedures are not the same and should not be transferred ... at least without some caveat to apply standards as "minimum requirements" when deploying fiber optic connections in field services.

Ideally, cleaning standards should be one and the same. However, production operations are always concerned with 'costs' * and "best practice" is to clean each connection every time it is "opened"...and this includes all jumpers and factory equipment.

* White Paper; September 2015 *Fiber Systems International.*
" How important is it to keep fiberoptic connections clean?"

Anatomy of a Standard…look outside our industry to understand the concepts and precepts!

SAE J 429 Bolt Standard:
(likely you are aware of it!)

Grade	Size Range	Tensile Strength min PSI	Minimum elongation %-	Material Hardness
1	¼"-1 ½"	36,000	18%	B7 to B100**
2	¼"-1 ½"	57,000	18%	B80 to B100**
5	¼"-1 ½"	92,000	14%	C19 to C30
8	¼"-1 ½"	130,000	12%	C33 to C39

Where is it used?

Grade	Possible Application
1	Toy Assembly
2	Light weight lawn mower assemblies
5	Medium weight assembly such as riding lawn mower
8	Heavy Duty assembly such as race car suspension

SAE J 429 is "applications specific"

"What if"…there were a fiber optic standard for inspection and precision cleaning that was "applications specific"?

What is meant by "applications specific"***

- ✓ The act of applying to a particular purpose or use relevance or value of practical applications
- ✓ Diligent effort or concentration. A job requiring specific application or detail.
- ✓ (Logic) the process of determining the value of a function for a given argument (COLLINS DICTIONARY)

*** Collins Dictionaries

WHAT IS THE VALUE OF A "STANDARD"?

Input from the brightest minds in our Industry

- **WITHOUT THEM: NO GUIDE LINES.**

- **<u>An Answer</u>: Create your own using the genius of IEC, TIA, Telcordia ... as a baseline.**

- **<u>Another</u>: you can input to standards committees.**

- **Urge your trade organization to implement annual or 'rolling standards' that coincide with technology ... not just every ten years.**

Eliminate Standards?

Absolutely not!

THINK OTHER WAYS:

Understand that existing standards are "minimum requirements" and not "best practice". [1.]

IT'S NOT THAT THE STANDARDS ARE 'WRONG'.

By the nature of how they are written, standards require much work and by the time they are published fiber optic transmission advances outdate them. Standards often result in a commercial venture. For this reason both are "incomplete" and a "compromise" of understandings of the science of soils, sciences of cleaning, and, the essential need to commercialize.

WHAT WE KNOW ABOUT 'FIBER OPTIC END FACE GEOMETRY' IS REALLY NOTHING NEW!

Logically, obviously: these structures are not two-dimensions in an x-y axis; but rather three-dimensions with an x-y-z axis. Likewise contamination that may be present on a fiber optic connector is infinite is type and wide ranging in location on these surfaces. Existing standards are based on 'easy to remove' contamination.

A higher standard works from "worst case to best practice. For these reasons you may want to create your own "Best Practice Standard" that is applications specific to individual jobs and locations.

Type	Anticipated soil	Capacity	Connector Type	To Do:
Data Center	Dust	100Gb+	SC/MT-Type	Train JB's Crew
FTTx Dry	Dust, Sand	1Gb	SC/LC	Order 1.25 swabs
FTTx Humid	Rain, mud	1Gb+	SC/LC	Rent recording scope

This matrix is a suggestion of how to approach an internal standard.

The Five Zone Three Dimensional View

To properly precision clean and inspect...
"Look Outside the Box"
> ➤ Consider total 'connector geometry'

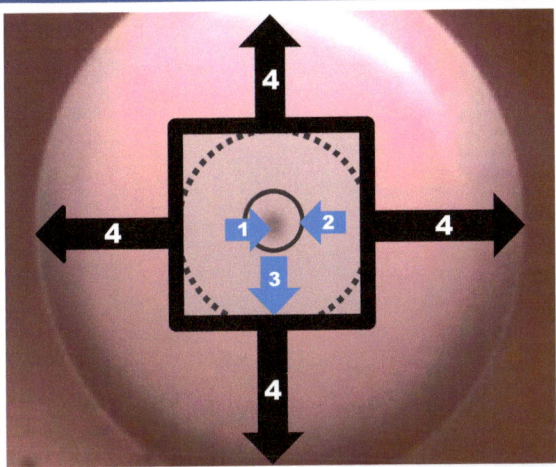

"Why"

The 2nd aspect of precision cleaning and inspection is to understand *"where" Is there potential for contamination.* For a fiber optic connection this means all parts, segments, or "geometry" of the connector become potential "soil points".

An advance over all existing standards: consider the end face as it is: *a three dimensional structure.*

There is a Zone-4 that extends outside the "field-of-view of the 200/400x inspection scope and a Zone-5 that cannot be seen with these instruments. "Zone-5" is not just the "vertical ferrule".

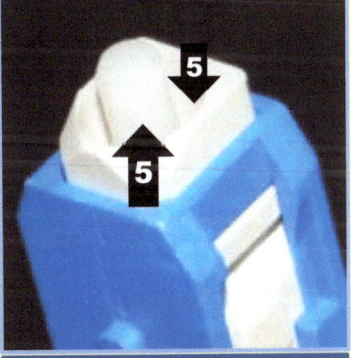

All parts of the connector can accumulate debris.
_____
Because of the three dimensional aspect of all things, a "search for best practice" means existing products and procedures may not be adequate.

A NEW VIEW:
there is more than an 'end face'

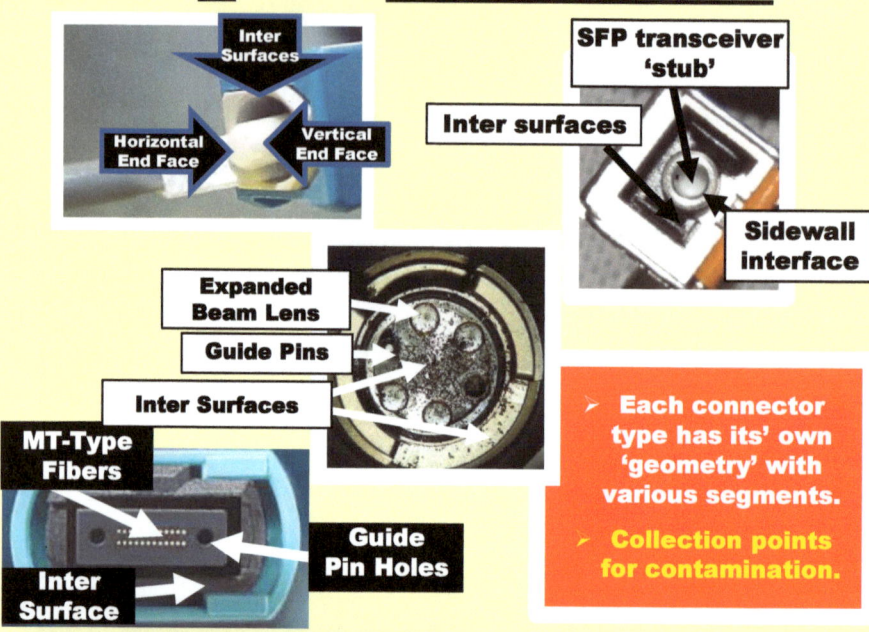

Inter Surfaces

Horizontal End Face

Vertical End Face

SFP transceiver 'stub'

Inter surfaces

Sidewall interface

Expanded Beam Lens

Guide Pins

Inter Surfaces

MT-Type Fibers

Inter Surface

Guide Pin Holes

> ➤ **Each connector type has its' own 'geometry' with various segments.**
> ➤ **Collection points for contamination.**

There are multiple components to every fiber optic end face as well as the areas that surround it all crating an actual connector assembly. While this is "obvious", less so is the actual interaction once inspection and cleaning is initiated.

All fiber optic connections have a three-dimensional construction with various segments that can either accumulate contamination from a work site …or… debris from a cleaning procedure. Yes, this is also "obvious" but not so much as far as a contemporary and practical cleaning and inspection process is considered.

The interactions between the connector's construction, a cleaning process and the actual contamination are integral in our search for best practice. Remember, there are three aspects to cleaning…anything. For a fiber optic connection it's critical to always understand that the small surfaces require special attention. While the fundamentals of cleaning have been "around" for more than 5,000 years…applying them to a fiber optic connection is new science when considered in regard to all existing standards.

Of course, we live in 3D.

Contamination is not two dimensional.

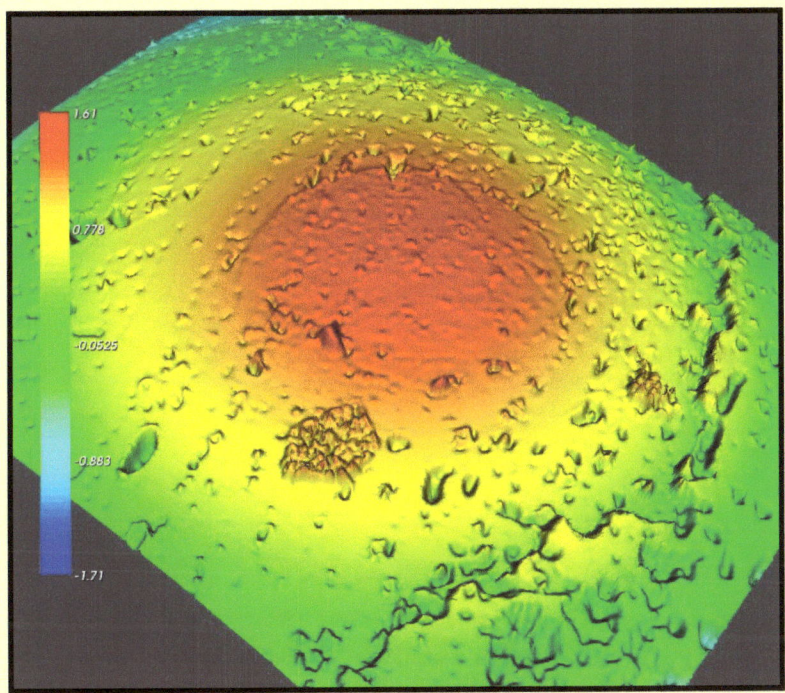

Is a two-dimensional view a Best Practice?

To properly clean and inspect a fiber optic connection **consider cleaning procedures that remove both the "height" and "diameter" of soils.**

IEC 61300-3-35 and all standards, as well as "automatic detection" systems are based on diameter of contamination in juxtaposition to the core.

In this interferometer reading (as well as more than 20 others) the height of the contamination equals or exceeds the diameter.

Among the implications are that height can create a stand-off between end face, damage the surfaces, or, in the case of an ineffective cleaning process remain as an "artifact" that *may have been removable.*

Microscopic Sizes

It takes just ONE piece of debris to *completely* block the fiber core.

Remember: the "core" is only 9 microns in diameter on a single mode fiber and 62.5 on multimode fiber

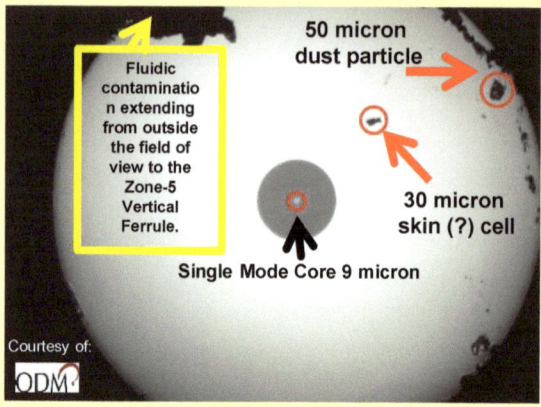

Fluidic contamination extending from outside the field of view to the Zone-5 Vertical Ferrule.

50 micron dust particle

30 micron skin (?) cell

Single Mode Core 9 micron

Courtesy of: ODM

IF THE CLEANING PROCESS IS NOT ABLE TO REMOVE THIS CONTAMINATION:

1.) The signal will fail

2.) You'll clean…and clean and give up or change the jumper or send the board back to the OEM!

3.) You might assume the residue is an acceptable IEC "artifact" and really have a problem!

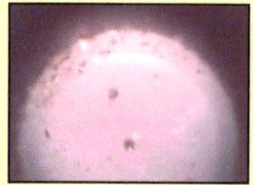

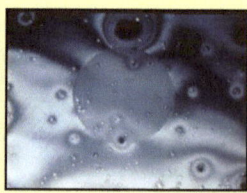

Contamination is present in many "flavors"! "Best Practice" understands facts-of-life!

Transfer of Fluid and Other Debris

Debris transfers when connections mated

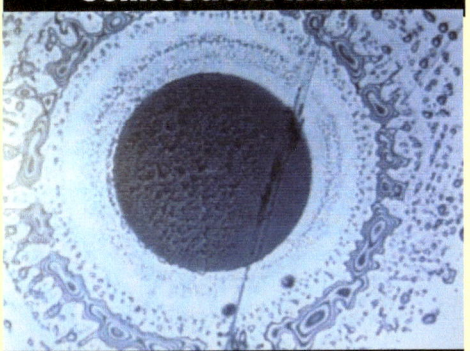

This unusual picture shows a concentric transfer of a compressed fluid transferred from one jumper to the other: further confirmation fluids have "height".

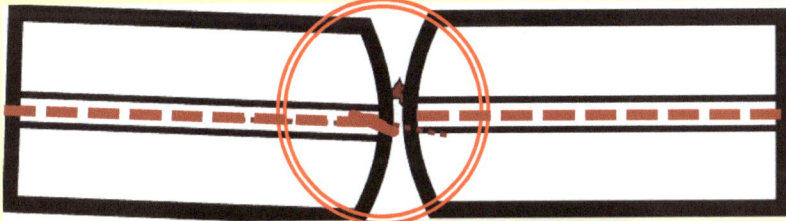

Dry debris trapped between the connection end face can create a "stand-off" and/or fiber misalignment

Remember the 3 things?

1st: What (connector type) am I cleaning?

2nd: What & Where is the contamination?

3rd: What process, tool, or device will successfully remove (any and all) contamination?

How do you know if the connection is clean?

- **What test equipment do I use?**

1. **Fiber Identifier?**
2. **Visual Fault Locator?**
3. **Db loss test set?** (power meter & light source)
4. **OTDR?**
5. **Video Scope?**

- **Do I need all this?**
- **How can I budget it?**
- **What works; what does not?**
- **Why?**

The "real question": *What do they do?*

Short course on test and measurement to decide:
"Which one(s) do I really need?"

Q: What is a Fiber Identifier

A: A Live Fiber Trace and Tone kit enables the technician to identify available fibers in the of the FTTx network without the disruption of existing subscriber services.

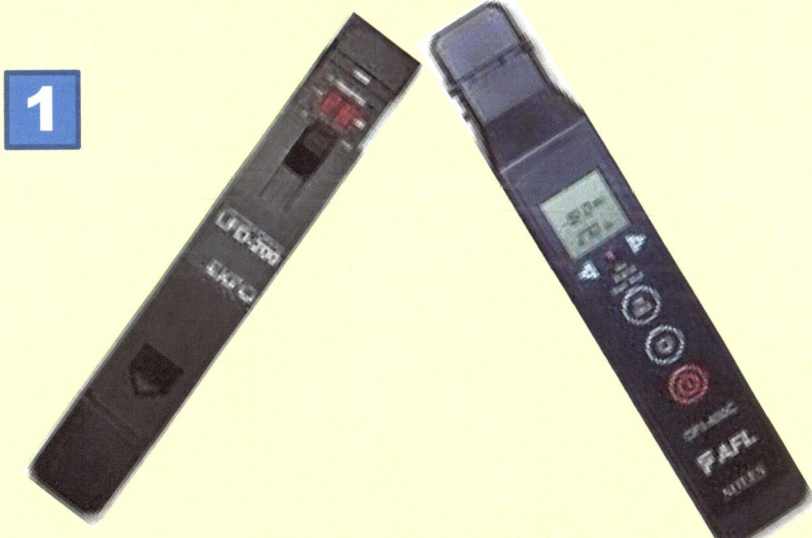

1

A fiber identifier will not determine that a connection end face is clean.

Key Thought: Just because there is a "live fiber" doesn't not mean the end face is actually clean!

Q: What is a Visual Fault Locator

A: A visual fault locator uses a high power visible laser designed to locate and identify faults in fiber optic cables and breaks in jumper cables, patch panels and other cable splice areas. They are typically effective up to 7km.

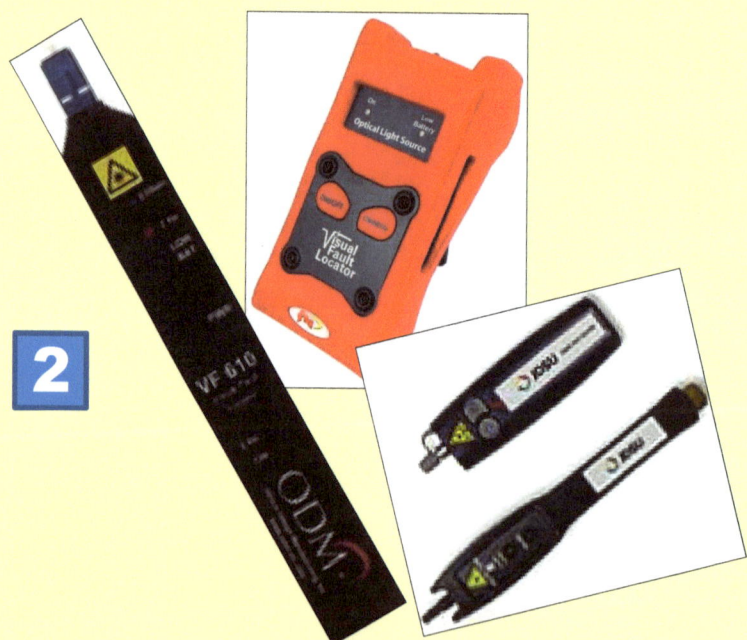

- These devices are an *inexpensive way* to locate sharp bends, some damage and conduct end to end continuity tests.

- *A visual fault locator will not determine if an end face is clean.*

- The device will not provide an actual measurement of loss or gain.

dB Loss Test Sets

Q: What does a loss set test do? **3**

A: It provides insertion loss measurements.

How much light is lost on a certain fiber run when compared to a reference value.

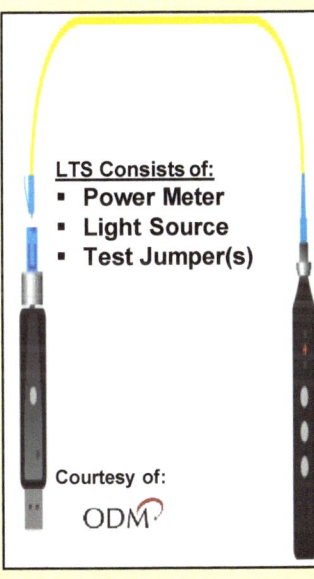

LTS Consists of:
- **Power Meter**
- **Light Source**
- **Test Jumper(s)**

Courtesy of:

ODM

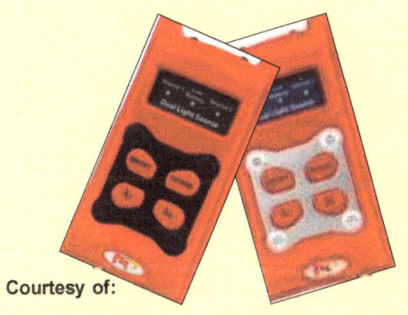

Courtesy of:

LOSS TEST SETS DO NOT PROVIDE:

Fiber endface images

Return loss values

An analysis: "Is a fiber actually clean"

These instruments simply provide a way to perform a logarithmic equation comparing two power levels.

$$L_{dB} = 10 \log_{10} \left(\frac{P_1}{P_0} \right)$$

It's much faster and more accurate than trying to remember algebra or calculus or geometry or whatever !!!

Optical Time Domain Reflectometers

Q: What does an **OTDR** do?

A: An OTDR provides a graph of a fiber optic cable, showing distance to faults and connectors, loss per km, and the length of the fiber run. The instrument will verify splice loss. Most commonly the OTDR will create an image of a fiber optic "run" at time of installation for comparison at another time. They are most effective "mapping" cable runs greater than 250 meters.   The OTDR also measures "reflectance". The OTDR will not measure insertion loss.

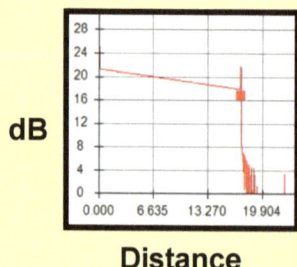

dB

Distance

They are best suited for long-distance fiber testing and measuring reflectance of connectors.

OTDRs are not as simple to use as a Loss Test Set, Visual Fault Locator, or, Inspection Scope.

The equipment is expensive...and may be rented or leased for occasional applications.

OTDRs do not provide:

-Fiber endface images

-Accurate dB loss values at short distances (<1km)

Connector Reflectance

What causes high levels of reflectance? Reflectance, optical return loss, or "back reflection" of a connection is the amount of light that is reflected back up the fiber toward its source. There is also a Fresnel reflection caused by the light going through the change in index of refraction at an interface. Reflectance is primarily a problem with connectors but may also affect mechanical splices using an index matching gel to prevent reflectance.

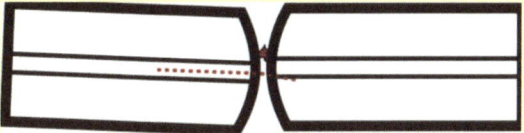

Air gaps, debris, and misaligned cores.

Since reflectance also can be the result from improperly cleaned end faces, or misalignment, the OTDRs may provide a way to identify dirty or faulty connectors.

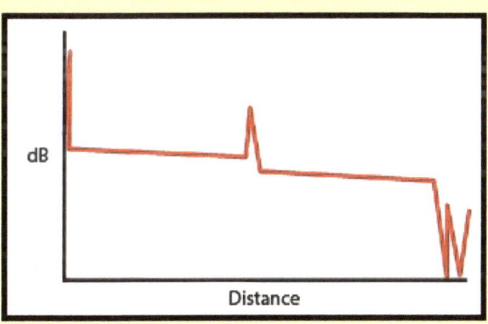

Using an OTDR, Reflection values can indicate a contaminated connection.

Measurable Effect of Soils Using an OTDR

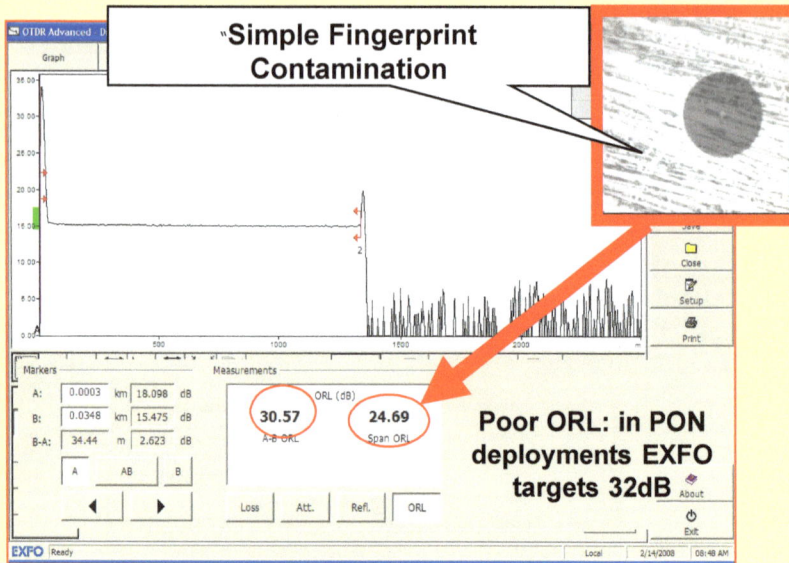

"Simple Fingerprint Contamination

Poor ORL: in PON deployments EXFO targets 32dB

APC jumper connected to ONT from drop terminal in ILEC/FTTH. Large reflection indicated dirty connector. ONT was returned to OEM

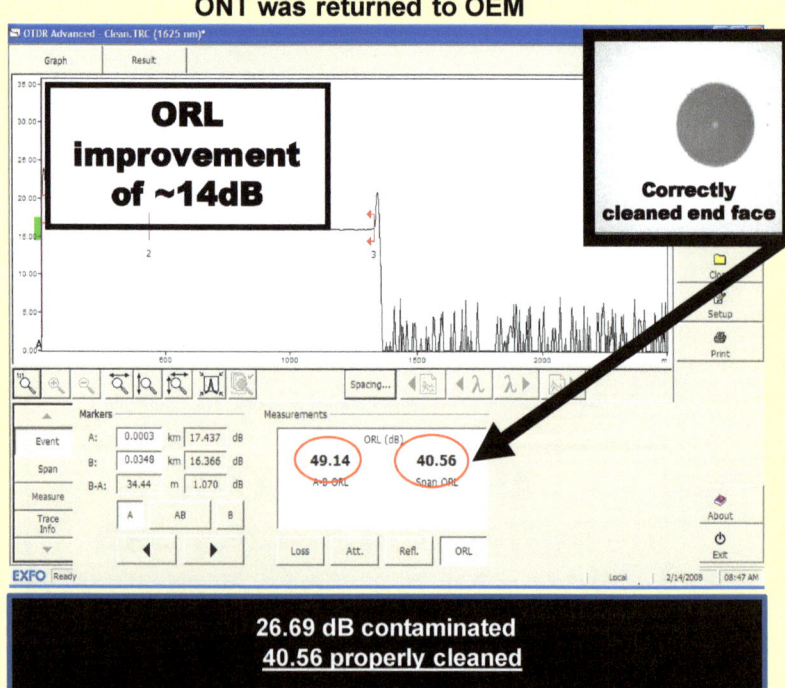

ORL improvement of ~14dB

Correctly cleaned end face

26.69 dB contaminated
40.56 properly cleaned

Courtesy of: **13.84 dB Improvement**

EXFO
EXPERTISE REACHING OUT

What is Inspection?

Q: What am I looking at when I use an inspection scope?

A: A highly magnified image of some segment of the "horizontal end face ferrule"

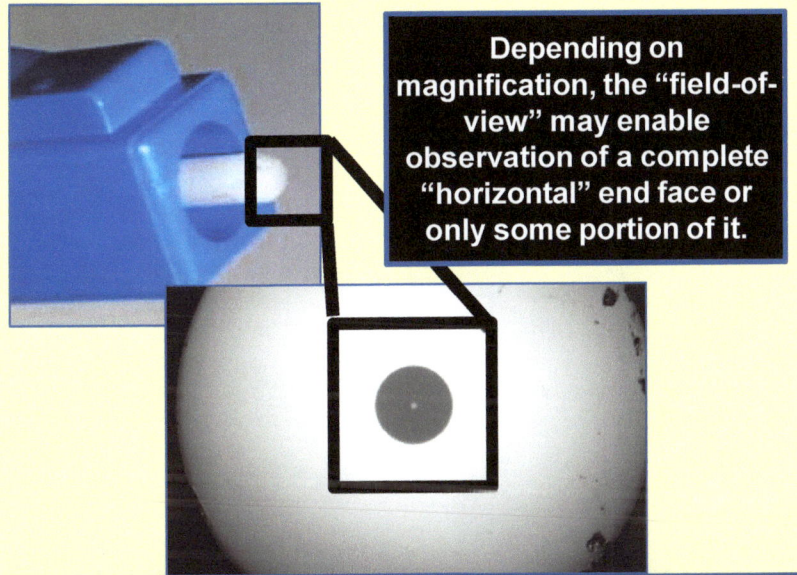

> Depending on magnification, the "field-of-view" may enable observation of a complete "horizontal" end face or only some portion of it.

WHAT IS "FIELD OF VIEW" AND WHY IS IT IMPORTANT?

Since we have established that contamination is present in many forms, field of view is the actual amount of the end face seen through a inspection device. Fluids outside the field-of-view can transfer in the time of post cleaning and inspection: this is a critical concern in the search for best practice.

There is a tradeoff as field of view is increased. This is typically done by decreasing magnification. By decreasing magnification the "resolution" (ability to actually discern contamination) is reduced. Knowing if contamination is real or just a defect is a critical function of inspection. Ask the supplier about "field of view" and "resolution values" for the instrument. Both are important.

What kinds of inspection are available?

What's It Do?   **What Do You Need?**

Class 1: Direct View and Loupes

CAUTION: THESE MAGNIFICATION DEVICES SHOULD NEVER BE USED TO VIEW A CABLE THAT COULD HAVE AN ACTIVE LASER SIGNAL.

Direct view magnifying scopes are adequate for jumper certification on a production line or prior to being placed in service.

A loupe can identify contaminates on the "vertical ferrule" or other areas of the connector as well as view certain expanded beam connectors.

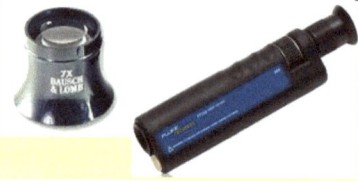

Class 2: Video Inspection

With a typical inspection range from ~130-400x. the higher the magnification the smaller the area actually seen. Lower magnification requires higher resolution.

Always select highest resolution when choosing low magnification and widest field of view. These devices are the "work horse" instruments. They have many features…select the ones you need. Plan for the future; avoid frills

WHERE ARE THEY USEFUL?

A loupe can inspect the outer surfaces of any connection. They are especially useful for certain expanded beam and "military style" connectors and to view Zone 5.

A hand held direct view scope is used to check jumpers in a QC function: a.) you just received 199 jumpers and want to check before you install for failures. b.) You have removed jumpers and want to store them for future use…check for damage, etc.

AVAILABLE WITHIN A WIDE RANGE, CONSIDER FEATURE TRADE-OFFS TO ASSURE EVERYONE HAS ONE!

There are highly effective "go/no-go" devices as well as more sophisticated units with variable view and recording capability.

I recommend the newer designs with wide field of view and high resolution. Check with your supplier and ask about features and their advances. Each has a story and rich R&D heritage.

What kinds of inspection are available?

What's It Do?   **What Do You Need?**

Class 3: Pass/Fail Video Inspection with Recording

This generation of instrumentation is based on standards traceable to IEC 61300-3-35 for diameter and location of contamination.

Some are self-contained while others work with a laptop or other device. Most record and store data images. Within limitations, most can be programmed to detect various types of contamination and debris. As of this writing (9-2015) none have the ability to detect height. Inquire about resolution and ability to discern a "defect/artifact" from actual contamination.

Since these instruments are more expensive, inquire about rentals for specific deployments that require their use.

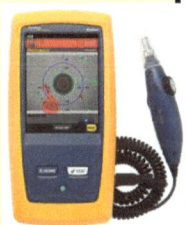

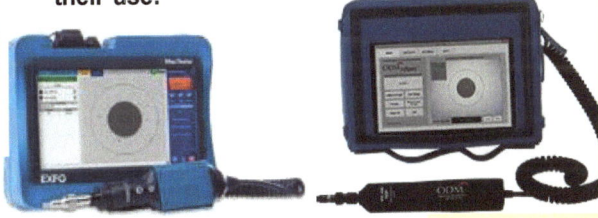

IF YOUR CLIENT DEMANDS
ADHERENCE TO
IEC 61300-3-35
THESE ARE MANDATORY.

On a day-to-day basis, there are numerous highly-effective Class-2 instruments.

A digital camera and spreadsheet may also work! Ask your client which you can use to prove your work.

Inspection scopes with a wide Field-of-View provide the best information

Wide F.O.V. for Best Results...plus exceptional resolution to determine the type of contamination.

Inspect and clean all the way to ferrule edges for best assurance of fiber cleanliness.

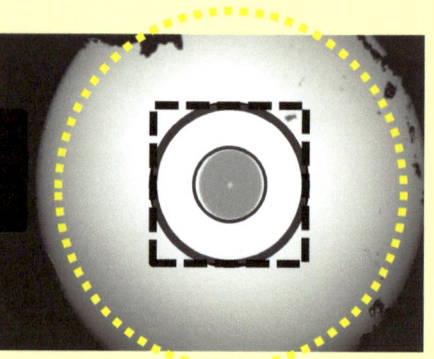

UNDERSTAND THE NATURE OF CONTAMINATION.

Something "dry" tends to remain in place.

Something that is "fluid" will move and transfer...over the "horizontal ferrule"... to the "vertical ferrule" and/or become lodged in the alignment sleeve and other parts of the connector geometry.

The field-of-view of even the most contemporary 125-150x "high res" video scope does not see the complete end face. The area of the yellow dotted line approximates the current limit. The area depicted by the black lines is the approximate viewing limit of a 200-400x inspection device.

Therefore a cleaning procedure that confidently removes all contamination from the end face without fouling the areas of Zone-5 is critical to the "Search for Best Practice".

Which one works best?

Amazing features for the future...which is now!

These instruments have grown with the Industry. Select features that match your business style.

A contemporary video scope is compact with a high resolution screen and wide field-of-view to see details of contamination.

There are "wireless' units that project on iPhone® or Android®

FIBER TEST EQUIPMENT CONCLUSIONS

Use an inspection scope every time a jumper is inserted into a patch panel or connection is made. Don't take a chance: clean all jumpers (even new ones) before they are installed. There are simply too many variables to "trust" anything other than your own work.

This is the ONLY WAY to be sure connectors are clean.

Power meter and light source (Loss Test Set) provides the simplest solution for identifying fiber optic cable viability...the "run" is (minimally) working!

OTDRs are most useful for long distance links and require more technical knowledge to operate and synthesize results. The OTDR tests the integrity of a fiber optic "run" and can verify splice loss, measure length and find other faults. It many not *conclusively* identify dirty connectors.

Always remember: "Just because you have 'light on a meter'...does not mean the connection is actually clean."

1ˢᵗ: What (connector type) am I cleaning?

2ⁿᵈ: What & Where is the contamination?

3ʳᵈ: What process, tool, or device will successfully remove (any and all) contamination?

Situational Awareness

The essence of self standardization and "best practice"

<u>Know Before You Go to Your Work Site:</u>

1. What kind of connectors will I be working with, and how many connectors are there?

2. Where are these connectors physically located, and how accessible are they?

3. To what of contaminates have these connectors been subjected?

4. *Is it "dusty in there"?*

5. *It is "moist and humid and muddy out there"?*

6. *Who called to ask? Where is the log from the last time?*

<u>Recommended Check List for the Work Site:</u>

1. Video Inspection

2. Cleaning products
   that match anticipated contamination

3. Have I been trained and updated?

4. Do we have an applications log?

One of the most valuable things you can have is a "flight log". Race cars and airliners have one and you should too!

Make notes of this job…for the next job and refer to them from the first time you were there….

Courtesy of:

MicroCare

50

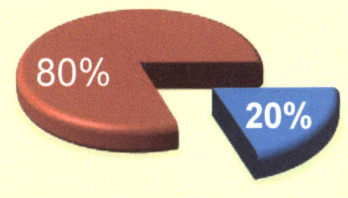

Cleaning Effectiveness:
Best Practice or
Does Pareto's Law Apply?

Named after its proposer Vilfredo Federico Damaso Pareto (1848-1923), French-born Italian engineer and a founder of welfare economics, it's also called 80/20 principle, Pareto's Law, or principle of imbalance.

From time to time you will hear it said that "...80% of contamination is dust"...easy to remove...stays in place...not all that great a concern! Some commercial tools are designed only to clean "dust"!

Pareto's observation was a large number of factors contribute to a majority result: this majority (about 80 percent) was due to the contributions of a minority (about 20 percent) of the same factors. In some contemporary businesses investigation does suggest that about 80 percent of the sales of a firm are generated by 20 percent of its customers, 80 percent of the inventory value is tied up in 20 percent of the items, 80 percent of problems are caused by 20 percent of reasons.

It is however a heuristics principle, and has not been proven as a scientific law. A heuristics principle is one that uses a mental shortcut for complex decisions! Neither is useful when considering precision cleaning and inspection! Furthermore, the "80-20 Law" has two sides to it: while one claims that 80% of contaminants is "dust" and somewhat simple to remove, the other is that 20% of the contaminants can be something else that is 80% of the overall problem!

When the cleaning and inspection procedure works in "Worst Case"...it becomes "Best Practice". Please turn the page and see studies that support this thesis.

One is by Cisco® and goes back to about 2005. Here a complex group of contaminants were presented that were far more difficult to remove than those in IEC61300-3-35. In December-2014 I expanded this test and compared various cleaning devices in an attempt to achieve "first time cleaning". This is significant because IEC61300-3-35 and all other standards suggest up to five times.

There is an important old adage: "Do It Right The First Time".

**After you clean...you must know...
is the surface actually clean.**

"Worst case" leads to Best Practice.

This is not a "new idea" !

IN 2006 CISCO® ISSUED AN INTERNAL STANDARD FOR CLEANING A FIBER OPTIC CONNECTION.

It was an important document that tested widely varied debris and contamination from the 1998/2008 IEC 61300-3-35 standard.

These included:

1.) Vegetable Oil
2.) Metal Shards
3.) Graphite
4.) "Duffy Dryer Lint"
5.) Simethicone
6.) Arizona Road Dust
7.) Dried water
8.) Dried 99.9% IPA

Graphite, Metal Shards Dried Water residues and Simethicone tuned out to be challenges!

Each contaminant was cleaned ten times using any specific cleaning tool, process or device.

A successful result was one that removed the debris <u>the first time</u> over the complete end face and repeated ten successive times

A copy of cleaning tests against these contamination types is available on request.

Courtesy of:

Chemtronics

Why is First Time Cleaning Important?

There is the logical answer: *"it's a time saver" ... which it is*.

The other consideration is if the process cleans the first time, through a wide range of contamination...it becomes a strong candidate for a "Best Practice".

Manufacturers design to standards, or, should a "best practice" define them? As you have seen and will continue to learn, a standard is not always a "best practice". Why is this the case? Largely because of the time it takes to write a standard for a rapidly evolving technology and then the time between to update it.

Not all cleaning processes and products clean the first time.

Why? Perhaps because IEC 61300-3-35 states that *cleaning up to five times* is *the standard.*

In December-2014 I updated the "Cisco-Series" using popular contemporary cleaning tools. There was "dry debris", "fluidic contamination" and "combinations of the two" evaluated. The goal of the experiment was to use each device one time in three separate processes. There were no "retakes" .

1.) IBC® Tool

2.) ClePen Tool™

3.) FerruleMate®

4.) FerrujleMate-2

5.) Sticklers® 2.5mm Swab tools

6.) ITW Chemtronics® 2.5mm Swab Tools

7.) QbE® Platforms

8.) Stickers® Clean Wipe Platform

9.) HandiMate™ Platform

1.) Vegetable Oil

2.) 10-30w SAE Engine Oil

3.) Arizona Road Dust

4.) Desert Dust from Afghanistan

5.) Jergens® hand lotion

6.) Vegetable Oil and Arizona Road Dust

7.) 10W-30 Engine Oil and Desert Dust

8.) Hand Lotion and Arizona Road Dust

The results from the 2006 Cisco® Series and the evaluation in December 2014 reconfirmed what has been field proven close to one million times over the last twelve or so years!

How to clean is not as much about the actual "product", but more about the "procedure". Tests in 2006 and then again in 2014 as well as throughout research in 2015 confirm what has been well known for many thousands of years. Adapting proven tenets is essential to fiber optic precision cleaning.

So far we have come to understand that inspection is an essential factor, but that existing inspection is not as effective as it could be. Some instruments do not see the entire end face and others are not adept at discerning all contamination types

This places even more emphasis on the need for a cleaning process that acts as a "safety net". One that works in a high percentage of the time in a "first time" brings us the end of our Search-for-Best-Practice.

IEC61300-3-35, TIA 455-240, SAE AIR6031 and numerous formal and informal documents and professional trainers speak about two procedures to clean a fiber optic connection. There are actually four.

1.) The first is defined as DRY PROCESS using a number of tools, swabs, and wiping materials

2.) The second process is a 'WET-TO-DRY' PROCESS, to be is used when the "dry process" does not work. On the following pages you will learn that this process may or may not actually dry the complete surface. Problematic here is cleaning without inspection may be as high as 50% of the time.

3.) Telcordia GR-2923-Core defines a third 'HYBRID TECHNIQUE': *use of a solvent and wiping material to assure the surface is actually dried and not contaminated with excessive liquid cleaner.*

4.) *There is a fourth process:* BLIND-CLEANING© in a "leap-of-faith" … without benefit of video inspection.

In a search for best practice evaluate all ways and means to arrive at the conclusion!

Fiber Optic Cleaning Product Options

With so many products, no wonder there is immense confusion and differing points of view! There are so many different products that the selection might cause one to ask: "Why has not one product risen above the rest?" The answer is that each has a significant weakness!

That weakness is not in the product itself, but how it is sold and used. The tests in December-2014 showed that just about each product could be made to perform better with one minor modification that does not require "going back to the drawing board".

IF you don't see your favorite tool here...it's because there was no more room on the page!

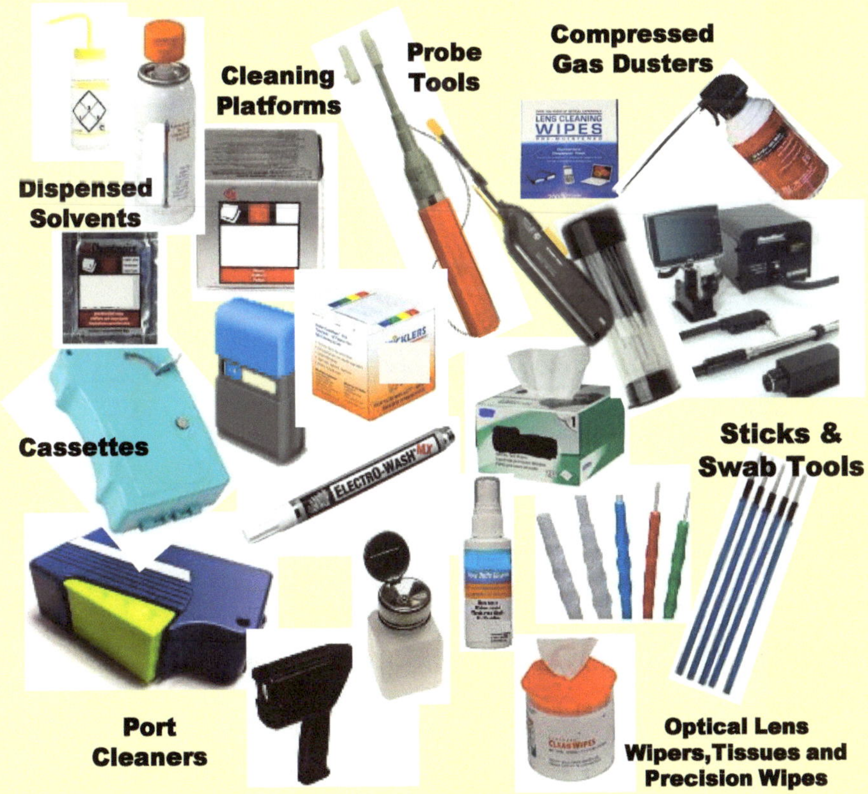

Dispensed Solvents

Cleaning Platforms

Probe Tools

Compressed Gas Dusters

Cassettes

Sticks & Swab Tools

Port Cleaners

Optical Lens Wipers, Tissues and Precision Wipes

Courtesy of:

Did you ever wonder?

Why are there so many
different cleaning products?

Doesn't it seem like only
one would be best?

Maybe cleaning is not about
"product choice" ... it's about a
procedure:
How a product is used.

To understand that...let's understand
more about cleaning products
themselves!

The Straight Story

What Works
What Does Not

...*most importantly*:

Why!

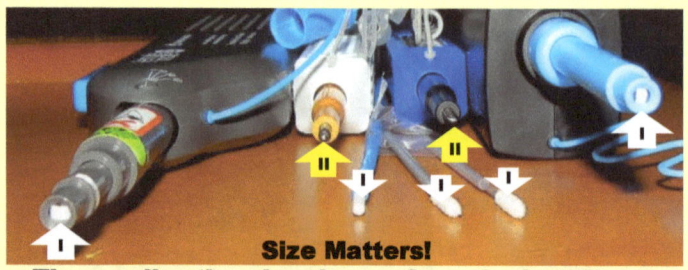

Size Matters!
The smaller the cleaning surface the less it can pick up dry debris or absorb fluids.

<u>THE SMALLER THE CLEANING SURFACE</u>:

1.) The less surface area it will clean
2.) The more difficult it is to absorb and remove debris or fluids
3.) The more times you will use the tool to clean
4.) The less likelihood there is for first time cleaning

<u>PRECISION CLEANING PRODUCTS AREN'T "ONE SIZE FITS ALL"</u>
This is an applications specific task.

Larger "tapes" and "swab tools" clean more of a 2.5mm surface and the complete 1.25mm end face.

Smaller "threads" are effective on the 1.25mm end face and less so on the 2.5mm.

Knowing this is information on <u>HOW</u> to use the tool. *Knowing this is not a reason NOT to use the tool.*

"IT'S SO MUCH HASSLE TO CLEAN".

Just this week there was a post on the FTTx blog with that quote. *I wondered to myself, 'what's the beef' with cleaning?* It is cleaning itself or is it the reality that once a connection is cleaned and scoped the surface isn't actually clean? Is it that once cleaned and inspected, the connection still causes a problem?

Maybe "the beef" is rolling another truck with an 'expert', over-time who just sent a text message that the 'problem' was just a dirty connection or the $250 invoice from a supplier who rejected a warranty claim because the $10,000 board you sent back wasn't defective...it just had a dirty connection!

Maybe "the beef" is that you just purchased 100 new cleaning tools and the 'problem' has not gone away. Maybe it's time to change from thinking it's better to have something that really works than something 'convenient' that fits in your shirt pocket!

What's "wrong" with 99.9% IPA?
ISOPROPYL ALCOHOL IS HYGROSCOPIC:

You've probably heard that before. It means it attracts moisture to itself. This means that the actual cleaning ability of the high purity solvent deteriorates, A 2003 study of 99.9% IPA showed that a measured container absorbed >3% moisture in less than fifteen minutes. "Retail-store IPA" may be 70-90% IPA with 30-10% moisture. Many times I have visited with technicians who reported and misunderstood: *"I was out of IPA so I went to the drug store"!*

A major factor effecting any and all cleaners is STORAGE. In these graphics pump bottles and Menda® storage bottles are shown to bring moisture (from where you are standing now) into any cleaner. Some cleaners can accept the moisture (which is humidity and itself may have micro contaminants) while others like isopropyl alcohol and acetone cannot. Fortunately, there are alternatives. Actually, an aerosol is a very clean delivery system: it's contents only "go" one way. Pens are another alternative.

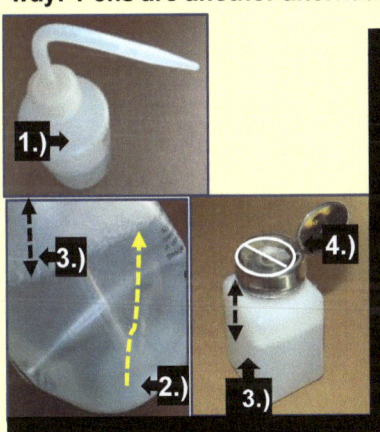

THE ANATOMY OF A CONTAINER

1.) As solvent is 'squeezed or pumped out"...

Upon release of the 'squeeze bottle' or depression of the Menda® Pump air is brought in to the solvent through the 2-way delivery tube. (2) Bubbles (follow yellow line) in the image far left are barely perceptible: a very real factor in solvent deterioration.

3.) "Head room" in the Menda® bottle & pumper further contaminates the solvent.

4.) IPA never should "hang around" in the basin: a.) Empty the bottle daily. b.) Do not cycle back to the original container!

Here's the real "problem": EVEN 99.9% IPA IS NOT A GOOD "UNIVERSAL CLEANER". *It performs well on polar (ionic) contamination such as body oil and salts, but no where near as well as precision hydrocarbons, many HFE7100 formulations or even new Aqueous cleaners on non-polar (non-ionic) contamination. Review all the contaminants in The Cisco® Series, my December-2014 Study, others noted on page 89 "References".*

Fact is that 99.9% IPA works on some types and not others. This is why IEC 61300-3-35, TIA 455-240, SAE AR-6031 and TelCordia GR-2023-Core all recommend NOT to use it as a fiber optic end face cleaner. 99.9% IPA is acceptable for Fusion Splice Prep...never perform a cross-application by end face cleaning with IPA. By the way: "drug store IPA" may be 5%, 10%, 15% 30% water: This is "The Straight Story".

Product Limitations: Wiping Materials

I believe that "precision cleaning" starts with what each of us already know about numerous cleaning products we use in every day life. How many times have you taken a facial tissue to wipe a runny nose and it just didn't do a "good enough job"!!!

Before the advent of ubiquitous plastic grocery bags, it was not uncommon to see one overloaded drop glass bottles in the parking lot...worse yet...when the paper bag got wet with rain or snow!

These "lessons of life" can and should be applied to the products we select to clean this amazing surface!

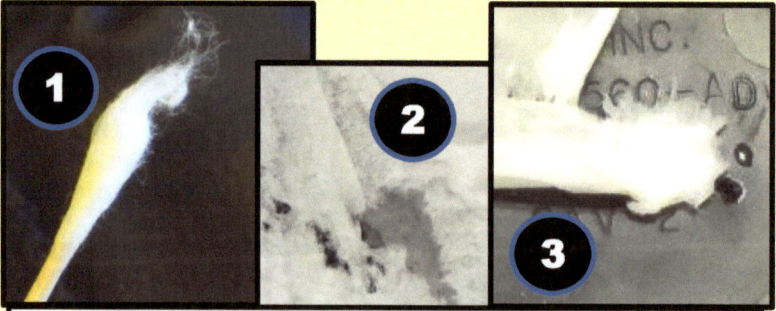

WHY NOT COTTON?
Cotton is an ideal material for "thread" and here you see cotton "threading". (1) This lint material can cross contaminate the port or be transferred to the end face. There is no such thing as a "lint-free cotton swab"...don't use it to precision clean fiber optics!

WHY NOT PAPER?
Paper is absorbent, but it does not have tensile strength. (2) When it tears it shreds.
(3) When it absorbs too much liquid, it disintegrates. If the product says "low lint" it is not _good enough_ to precision clean fiber optics!

What Works
What Does Not
...*most importantly*:
Why!
How to choose...

Let's face it! There are many reasons "why" each of us choose a product! Some like the store and have always purchased from the same distributor. Others buy whatever their best friend is selling! Some purchase based on "price"...cheapest or Porsche-mentality: the best! Some research and still make a mistake...others seem to get it "spot on". Making a buying decision is not easy.

On the following pages is are product matrix of various products. By nature of the fiber optic industry with so many advances coming so rapidly, it's a 'foolish-mission' to point in one direction.

THE POINT is to be aware of these factors when selecting precision cleaning products for fiber optics. Over the last ten years there have been and will continue to be advances in precision cleaning products. Many of these are based on "convenience" and it's not clear that convenience is actually "Best Practice".

Have you ever tried to remove a Philips® screw with a 'straight-blade' driver, a 10mm bolt with a 1/4" wrench, or, hammered a nail with a brick?

Precision Cleaning Fiber Optics is an Applications Specific Process that with the proper technique...is as easy as it is satisfyingly effective! The "hassle" is gone....

When reviewing "Strengths and Weaknesses" ...don't consider them as "knock outs".

It's likely none are "perfect". Seek Best Practice.

Product Limitations:
99.9% IPA, KimWipes® and Canned Air

Product	Limitations
Pumps and Squeeze Bottles	• **Uncontrolled environment** • **Squeeze draws in air causing cross contamination and product dilution**
Moistened Lens Grade Tissues	• **Wipe materials may not be optical grade** • **Solvents may not be optical grade**
Natural Fiber Products (Paper, cotton)	• **Lost fiber residue** • **Limited absorbency**
Compressed Gas Dusters	• **These are NOT "canned air"...it's a high performance gas.** • **Ineffective on OILY residues** • **Will not remove surface bonded debris** • **High velocity "all-way" containers may be useful for storm or water damage prior to precision cleaning**
Air Compressors	• **This is "canned air"...requires filtering to remove moisture and compressor lubricants.** • **Typical for production line operations** • **Although heavily filtered...may depart residues from the equipment.** • **Not typically used in field operations**

"Residues" are many things!
1.) Remnants from an ineffective cleaning procedure or paper from a wiper.
This does not mean "don't use them":
"know the limitations of the product and the process".

2.) A solvent that draws moisture to itself or does not completely dry is another "problem". A moisture residue is left if the wiping material cannot absorb the residual.

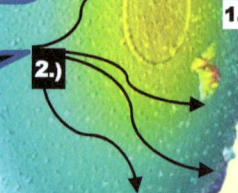

1.)

2.)

Courtesy of:
STICKLERS

Product Strengths and Limitations

Product	Strengths and Limitations
1. Isopropyl Alcohol 2. Precision Hydrocarbons 3. HFE-7100 4. "Aqueous Cleaners"	1. Hygroscopic, nature further diminishes limited cleaning ability, Flammable; Aroma. Has shipping restraints. 2. Highly Flammable, exceptional cleaning ability, aroma. May have some environmental limitations. Has shipping restraints for some containers. 3. Non-flammable, limited cleaning ability compared to hydrocarbons, increased environmental scrutiny. Aerosols may have shipping restraints. 4. Non-flammable. Must be actively dried. Excellent cleaning ability. EZ Shipping.
Pumps and Squeeze Bottles. Pens.	• Squeeze containers and pumps draw in air causing cross contamination • Pens are one-way. Make sure tips are clean
Moistened Lens Grade Tissues	• Wipe materials may not be optical grade. Read carefully...if the product cleans a microscope or binoculars...it's likely not right for fiber optics!
1. Natural Fiber Products (Paper, cotton) 2. Non-Woven and synthetic fabrics	1. Linting caused by shredding paper or cotton fibers 2. Stronger than paper but not all are as absorbent. Preferred over 100% or cotton products.

Seek 'trade-offs' that result in Best Practice for each installation. This is not as difficult as it sounds...there may be only 3-4 variables to consider in your lifetime!

Courtesy of:

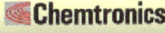

Product Strengths and Limitations

Product	Strengths and Limitations
Cleaning Platforms	<ul><li>Only clean jumper side or accessible from rear of equipment</li><li>Larger cleaning surfaces remove contamination away from point of contact</li><li>Larger containers some tend to be inconvenient</li><li>Disposable components, lowest costs per cleaning.</li></ul>
Probe Tools Sticks and Swab Tools.	<ul><li>Exceptional convenience cleans both jumper side and back plane</li><li>Smaller cleaning surfaces may not absorb as effectively</li><li>May clean alignment sleeves and other connector "geometry"</li><li>May require multiple cleaning passes</li></ul>
Cassette, Reel Cleaning Tools	<ul><li>Exceptional convenience, refillable, cost effective</li><li>Smaller cleaning surfaces not as effective on wide range of contamination</li><li>May require multiple cleaning passes</li></ul>
Mechanical Cleaning Devices	<ul><li>Larger systems more suitable for OEM production</li><li>Higher cost of acquisition</li><li>Inadequate solvent may leave residues that surface bond to the end face making contaminates more difficult to remove.</li></ul>
ALWAYS CHALLENGE THE SUPPLER... COMPARE AND CONTRAST	Obtain a sample...but do "them" a favor: tell them how it worked for you!

Important Note: every product has a "strength and weakness". Just knowing this is the 1st Step to "Best Practice".

How to Clean can be confusing!
The Essential Information

Is there really "this much" to all of this? Is taking this decision all "that" complicated? Most of us don't realize how we have conditioned ourselves though the years of cleaning product selection to achieve "best practice". We wash a floor and it's clean...if not we may do it again or select another cleaning product. Washing a floor is not as "mission critical" as can be precision cleaning a fiber optic connection!

Let's take a look at how we clean a fiber optic with "products" and specific "procedures". Please note the hyperlinks that will take you to YouTube® video of live demonstrations. For this record, these demonstrations have been performed "live" many thousands of times. The results are "repeatable" which is an important tenet of "best practice". It's one thing to do it one time...but when the same result returns thousands of times it is worthy of consideration.

These are the cleaning methods available in 2015

1. A DRY PROCESS using a number of tools, swabs, and wiping materials. IEC61300-3-35 and many other standards and trainers suggest this as "the first step" to clean.

2. A 'WET-TO-DRY' PROCESS which may or may not actually dry the surface. IEC 61300-3-35, other standards and trainers advocate this 'if dry cleaning doesn't work'. (If my boss didn't buy me a scope...the battery is dead...how do I know?!)

3. Telcordia GR-2923-Core defines a 'HYBRID TECHNIQUE': *use of a solvent and wiping material to assure the surface is actually dried and not contaminated with excessive fiber optic cleaner*

4. BLIND-CLEANING© "leap-of-faith" ... without benefit of video inspection. This "non-method" is still common and there are many reasons it exists. Yes, it is not "desirable" but as we search for best practice we have to consider "reality"!

Let's Study Each Procedure.
Be Open to Change.

DRY PROCESS USING TOOLS, SWABS, AND WIPING MATERIALS

❑Advantage: The Convenience of Tools

❑Disadvantage: Moves Dry Debris; may not remove it.

❑ Dry clean *fluidic contamination* to "dry-mop" the surface. "Best Practice" is to use the "Dry Cleaning Process" when the contamination is determined to be a fluid.

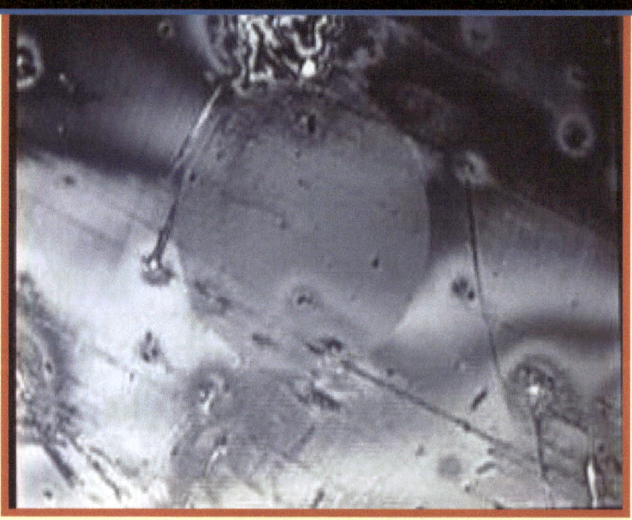

To see this demo please go to:

http//youtu.be/cJPbbNA11Jg

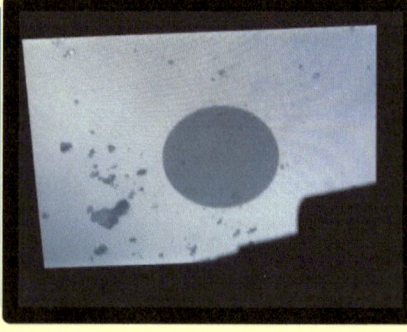

Another factor is Static Field Contamination from the Dry Cleaning Process

What is ESD? (Wikipedia, ESD Association, www.esdsystems.com)

- Electrostatic discharge (ESD) is the sudden flow of electricity between two electrically charged objects caused by contact, an electrical short, or dielectric breakdown.

- A buildup of static electricity can be caused by tribocharging or by electrostatic induction.

- The ESD occurs when differently-charged objects are brought close together or when the dielectric between them breaks down, often creating a visible spark.

<u>Methodology:</u>

In 2005 a study was conducted by ITW Chemtronics® regarding static field contamination.

In it the following observations were recorded. These are proven tenets, well documented in other Industry sectors such as production electronics. You have likely experienced them also as you touched a door knob when it was at times of low relative humidity!

> Static is best controlled by creating an increase in relative humidity. ie: by using a solvent.

> Static is not easily controlled by dry materials. Any treatment of these materials will leave contaminant on an end face.

> Static can be environmentally created by low humidity and temperature extremes. It is most commonly generated by a tribocharge.

What is a tribocharge? *Tribocharge simply means the generating of a static charge on two surfaces that come in contact and then separate with each other.*

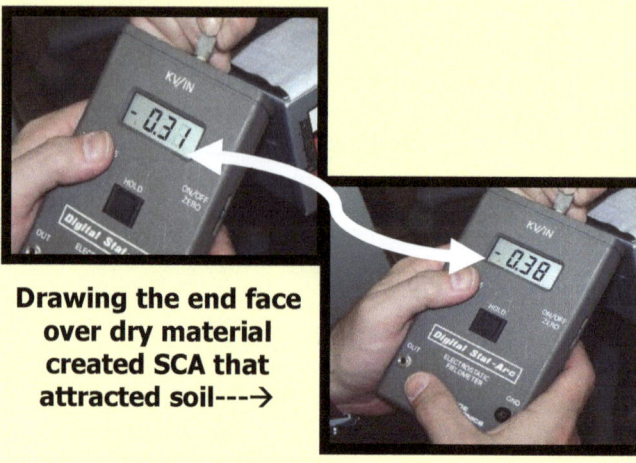

Drawing the end face over dry material created SCA that attracted soil---→

- **End face lightly drawn over wiping material 3 times**

- **A measurement of 0.31kv was obtained.**

- **Subsequent measurements ranging from 0.31kv, 0.34kv, 0.36kv, 0.38kv were obtained in the same session**

"Dry" testing also performed with
a.) "reel" cleaning tool,
b.) ferrule cleaning tool
c.) "lint-free" paper wiper *obtained like results*

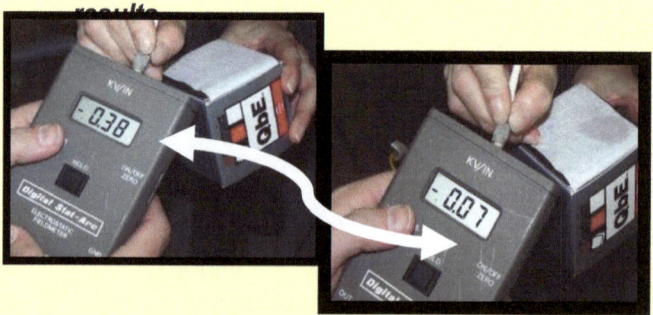

When the study was repeated using a moistened wiper...static field was significantly reduced.

Laboratory studies demonstrate lint particulate is easily attracted by relatively low Static Charge Accumulation readings

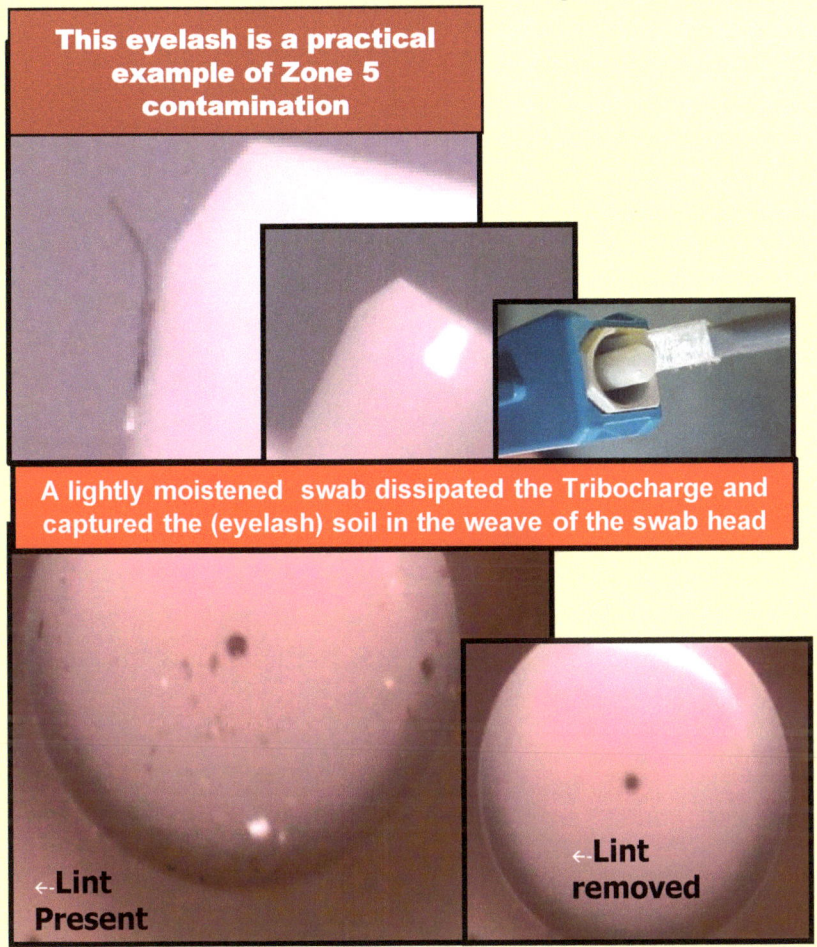

This eyelash is a practical example of Zone 5 contamination

A lightly moistened swab dissipated the Tribocharge and captured the (eyelash) soil in the weave of the swab head

←Lint Present

←Lint removed

HOW MUCH DOES AN 'EYELASH' INFLUENCE
A FIBER OPTIC TRANSMISSION?

The light contamination in the image immediately above this paragraph to   left can be more 'practical' than an eyelash…as well are more of a problem.

That light 'dusting" likely has a static bond to the end face. Depending on what it is, removal may be more of a problem that one can imagine. To remove this type requires a cleaner with a low surface tension (to actually get under the dust) and a wiper that will not fall apart during the process.

Courtesy of:
Chemtronics®

Typical Relationship Between Relative Humidity (RH) & Static Charge

- **Typical premises Relative Humidity levels are in the 25% to 40% range**
- **Relative Humidity 'outside' can be infinite day-to-day!**
- **Humidity levels vary with the room or outside ambient.**
- **Introduction of a cleaning fluid creates a medium for the static charge to 'dissipate'.**

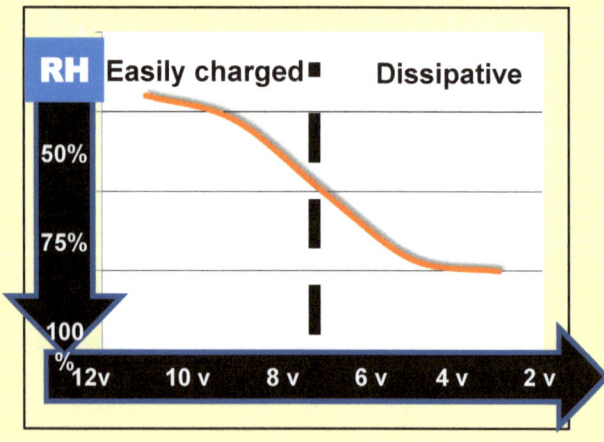

<u>THERE ARE SEVERAL WAYS TO CONTROL STATIC</u>:
- **The first is use of a grounding device. This is not functional for fiber optics as there is no 'path to ground' with a wrist or heel strap**
- **A production floor uses 'air ionization'. These devices are somewhat cumbersome and not practical for field applications**
- **Static dissipation...in effect...increasing relative humidity by use of a precision solvent is most appropriate for field service...as well as OEM applications.**

Courtesy of: **STICKLERS**

Dry Cleaning Conclusion
(View YouTube®...also try it yourself)

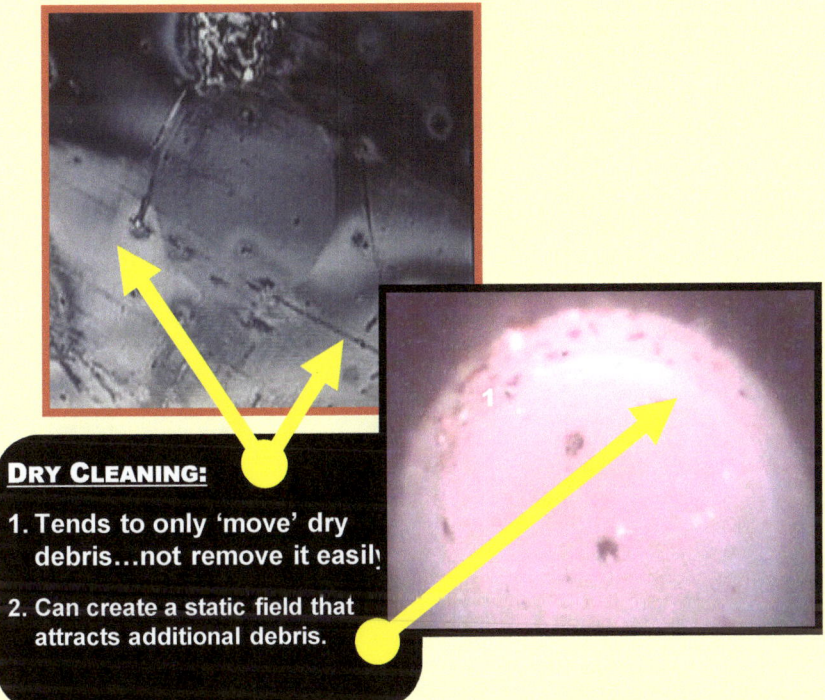

DRY CLEANING:

1. Tends to only 'move' dry debris...not remove it easily

2. Can create a static field that attracts additional debris.

Dry Cleaning is best as a "mopping action".

Think about drying your hands after they are wet ... a floor with a spill.

These are time-proven "sciences of cleaning" and "sciences of contamination" adapted to

Precision Cleaning a Fiber Optic Connection

Please request lab study on static field contamination

What's in a word?

I believe that every word has meaning! In the case of some, it's okay to be a little vague … maybe even desirable!

In those instances where we want a specific outcome we look for clear instructions. In this segment we are going to evaluate the words and instructions "wet-to-dry cleaning". To me, it's a little vague with interesting consequences.

First of all, give credit to those who proposed the idea of using a solvent, thus adding "wet" to the fiber optic cleaning process. _On the other side, what is missing is "how, when, where and to what degree" of solvent actually to use!_

"Wet-to-Dry" Cleaning

Everything in Moderation

    If you subscribe to the notion we exist in a three dimensional environment, then the idea of "wet-to-dry" has a literal meaning: 1.) Can the "wet" can be "dried" when in fact and reality: 2,) if not controlled and defined, excess cleaner can migrate throughout the various recesses of a fiber optic connector?

    In fact, simply saying "dry it" has some meaning … but the action may be more of a "wish, hope and leap of faith" than "Best Practice".

How does contamination "happen"

Condensation Contamination

The image was taken in Pittsburgh, PA in late February-2013.

The jumper was at (27F) ambient and then inserted into the video scope at ~70F inside ambient.

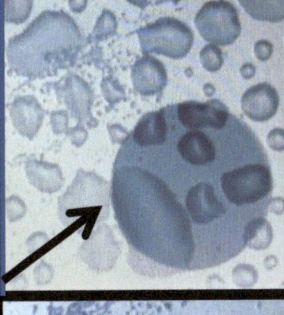

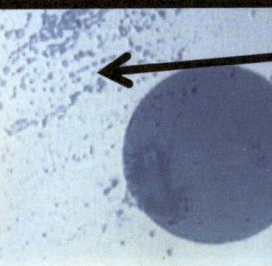

Although the condensation evaporated quickly (15-20 seconds) this residual contamination remained.

The residue surface bonded to the end face and was difficult to remove. **

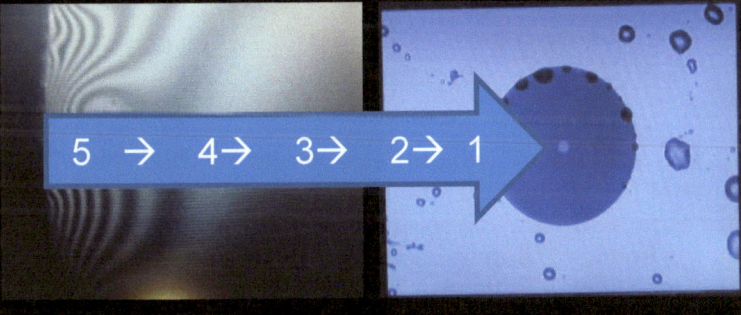

5 → 4→ 3→ 2→ 1

➢ Excessive cleaner may be trapped in Zone-5 or positioned outside the 'field of view'

➢ Don't let the term 'wet-to-dry' convey a sense of over-confidence

"Fluidic Contamination" on the outer limits of the "field of view" can transfer and migrate.

HOW OFTEN DOES THIS HAPPEN?

If you are 'blind cleaning' you may never know!

** Residual Contamination on Fiber Optic End Faces from Use of Polar Alcohols Paul Blair and Edward Forrest ITW Chemtronics June 2002 (rev:11/03)   White Paper (Discussion of surface bonding)

"Wet-to-Dry" Cleaning

"What's in a Word?"
What we say should e what we mean!

WHERE'S THE BEEF?

"Thanks" to Wendy's®!

VIDEO LAB
"Wet-to-Dry" Cleaning
https://youtu.be/qZTCutx2hsg:

Did you smile? When I saw the "small beef patty" in the center of the "huge bun" the entire assembly reminded me of a fiber optic end face! Sure, it's funny!   Recently, at one of my training seminars someone asked me" *"Where's the Beef"*...

Some may have seen this (highly repeatable) demonstration and if not, please check the YouTube®. This end face is cleaned with an IPA Wiper...and the result is "not so funny"!

The "field of view" is clean and dry. Automatic software says this is "good to go"...*except excessive solvent is trapped in Zone 5 and works its way to the center and Zone-1 ... sometime about the time you arrive back to your office!*

"WHERE'S THE BEEF?"   Words have meaning and in this case just saying "wet-to-dry" may not actually be clear enough. If nothing else when someone says "wet-to-dry" be aware and under-use the amount of cleaner. If the end face is still not clean...use a different cleaner...it may not be effective on the contaminant you are trying to remove!

The 3th Cleaning Option is Best Practice Precision Cleaning

"Hybrid" or Combination Cleaning Process

Often overlooked, this technique is noted in the most recent Telecommunications Industry Standard Telcordia® GR-2023-Core. This document was published as an update to IEC 61300-3-35 in Spring 2011. It is the most recent work of its kind.

The "GR" is written with a deeper understanding of the 'science of contamination' and 'the sciences of precision cleaning'. It notes three cleaning styles, adding one based on an important tenet: "contamination is attracted to moisture". The document does on to express concerns about over use of any cleaner and advises that 99.9% IPA is effective for fusion splice prep but to acceptable for end face cleaning. How many cleaning "procedures" have you done in your life when adding "something wet" made cleaning easier and better?

That "everyday science" is the essence of Best Practice cleaning of a fiber optic connection...especially if you are 'blind cleaning'. I strongly encourage and endorse inspection of each connection every time it is "opened". I also understand the reality that this is not possible and certainly not something that is 'enforceable'. What is more practical is placing a cleaning method in each work center, central office and work truck as the only approved cleaning procedure. By placing the right tools in craftsperson's hands...the job can be done right the first time...more times than not.

What is "precision cleaning" using a "hybrid technique"?

As it turns out, "Precision Cleaning" for "Best Practice is intuitive and easy to understand, train, learn and actually perform! The procedure combines science and craft...something that most technicians seem to innately understand, appreciate and enjoy.

Some may claim they are 'already doing this'. It's a good idea to make sure by placing the right products in working hands...making sure the local rep trains once and comes back a couple of times a year. Why? Old habits can be hard ones to change.

Depicted here are "products" you may have tried or are already using. This is not a "product discussion" that promotes a specific supplier: it's rather an "applications-specific technique that may have been used 1,000,000 times since 2001!

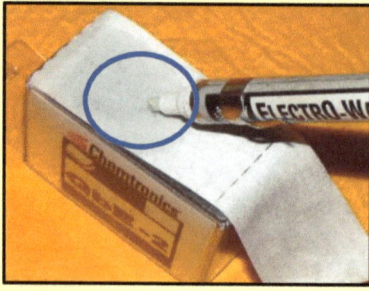

PART-1: How to avoid over saturation of the end face: start with a small amount.

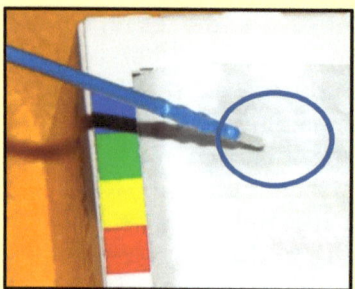

PART-2: Place the swab tool or probe into the "spot" of solvent. This transfer of precision cleaner is enough to clean a wide range of contamination...and will not "flood" the end face. Doing this "the first time" is "Best Practice" for any cleaning device.

What is "precision cleaning" using a "hybrid technique"?

As we discussed, swab tools and probes have small cleaning surfaces. Don't be too surprised of you must clean a few times. You will find that a lightly moistened probe or swab should clean 'faster' and better than using it "dry"...and chances are you will exceed the IEC standard of doing it five times...when the tools are lightly moistened.

Advanced technicians often prefer the 'inconvenience' of a cleaning platform and also will use one of the 'reel cleaners'. A larger cleaning surface removes more contamination away from the point of contact. Swabs and probe tools rotate the contamination and may require multiple actions to move debris away from the initial point of contact.

However...there is a bit of confusion: **what is the actual cleaning motion?**

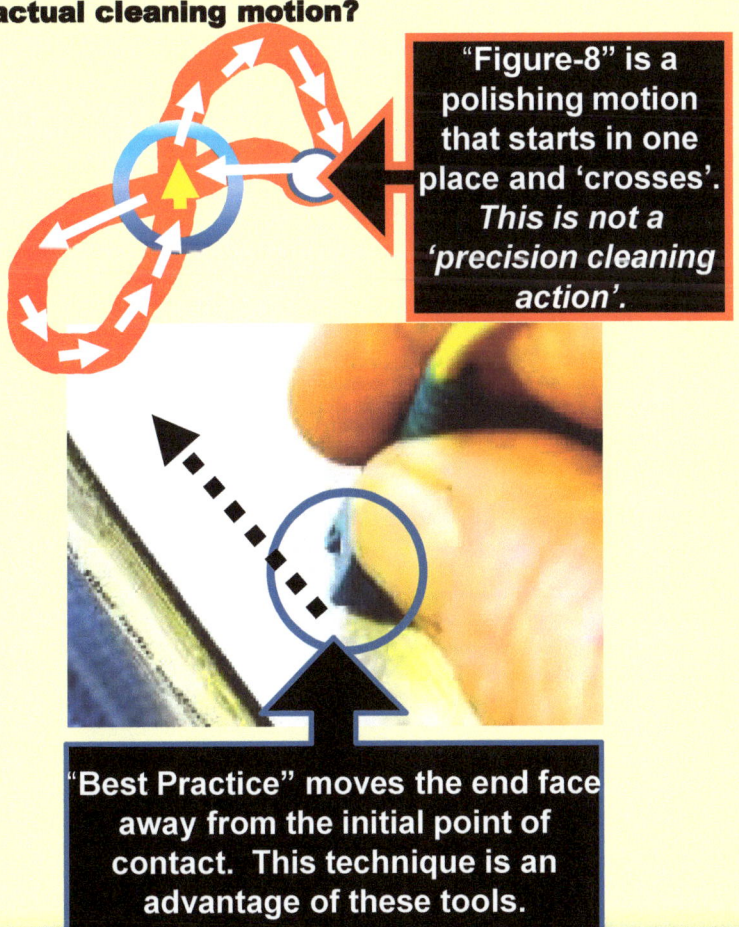

"Figure-8" is a polishing motion that starts in one place and 'crosses'. *This is not a 'precision cleaning action'.*

"Best Practice" moves the end face away from the initial point of contact. This technique is an advantage of these tools.

Video Lab Using Precision Cleaners and Wiping Devices in a Best Practice Process

VIDEO LAB "HYBRID" OR COMBINATION CLEANING
https://youtu.be/462arRGe7tA

Static Measurements *significantly reduced* using the Hybrid/combination process

There are three readings here: .03, .05, and .07... these lowered levels of static field do not attract dust

Match the contamination to the solvent type

Move the end face *away* from the initial point of contact so there is no recontamination

- **One product is not necessarily better than the other.**
- **Each has strengths and weaknesses.**
- **Knowing "procedure" leads to "Best Practice"**
- **Ask your rep to help...request factory involvement or seek SME for 'tough nut' problems!**

How to Clean can be confusing!
Eliminate the Confusion!

What Works!

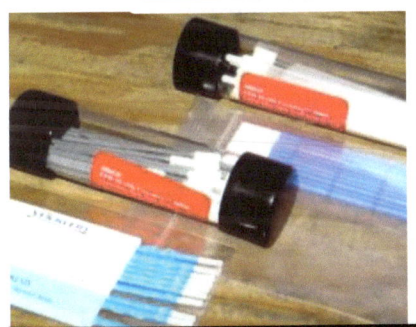
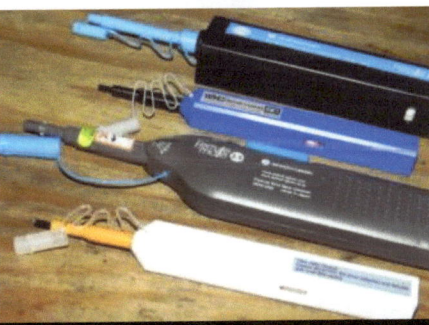

This group of products used with precision fiber optic solvents is "Best-Practice"

Which one? It is an 'applications specific" decision.
Like a Straight-Blade/Philips/Torx decision...your
cleaning kit may need all types to do the job right!

Ask your Factory rep or your distributor
to challenge the factory!

Be Prepared® ... those Brontobits are Coming!!!

Eliminate Confusion!

Understanding the Strengths and Weaknesses Does Not Always Mean One Product is Not as Good as Another.

Understanding is Best Practice!

THE FIRST TIME

1.) Always use a small amount of fiber optic cleaner that matches the contamination

2.) Move the end face away from the initial point of contact.

3.) Swab tools and probes may require multiple efforts...they rotate debris.

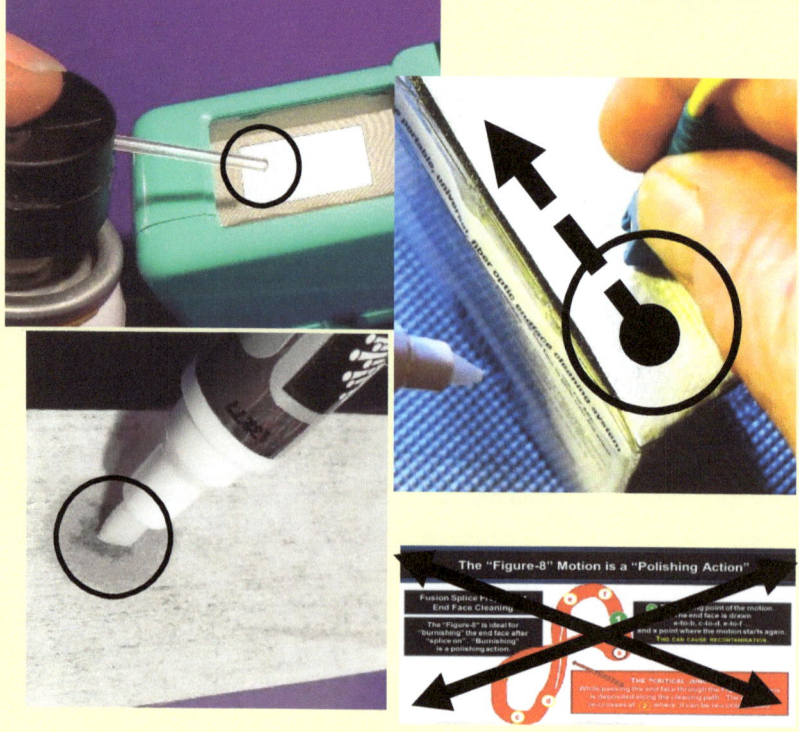

Contamination of all types is attracted to 'moisture'

BEST PRACTICE IS NOT TO "DRY CLEAN FIRST" AND THEN FOLLOW WITH A "WET-TO-DRY" ... IF THAT DOES NOT CLEAN. (IEC 61300-3-35, TIA 455-240, SAE AIR 6031, ETAL)

Best Practice is 1ST-TO-USE a small amount of fiber optic grade precision cleaner, clean room grade wiping materials, and a procedure that moves the debris away from the initial point of contact.

Most typically this simple process change results in first time cleaning and not multiple efforts...

How Does this Happen?

Case Study: March-2014
176 Fiber Optic Technicians were asked:

"How many have and use
a video scope?"

27 of 176 "working with fiber" had inspection "of some type" and/or used it regularly.

Conducted by Edward J. Forrest, Jr. at Iowa-Nebraska Telephone Association Annual Meeting: Des Moines. 2014

BICSI Regional Meeting, Charlotte, NC. :

only ½ of those who had inspection ... used it!

July-2015

Conducted by Edward J. Forrest, Jr.

"For the record"...I do not work for a cleaning company or one that sells video inspection! I am a researcher, trainer and author. I believe that inspection is important and realize, from years of being out there...there are "reasons" why not all connections can be inspected.

That is the reason for this book...to advance knowledge and this science. I will create a program for you, train your personnel or license one of my sessions with audio and video. Thank you for purchasing this book!

Conclusions: What does it all mean?

- Fiber Optic Connectivity is *(negatively or positively)* impacted by how it's cleaned

- Existing Standards are "base lines" and not "best practice". Consider 'internal standards' not only for your business but also for a specific installation or design.

- Selection of the right piece of test equipment to know if the connection is actually clean is clear: it's Video Inspection. Remember, if you don't have budget to "buy"...a rental is a fine idea to get you through!

- First Time Cleaning a fiber optic connection is easy once you understand the science. Remember: dry debris is attracted to the moisture of a fiber optic precision cleaner and some fluidic contamination "demands" a solvent with a wiper for break down and removal.

- Always use a high quality fiber optic grade cleaner and non-paper wiping materials. *Ask the supplier for details about products you may not understand. A distributor is an important resource...also...do not hesitate to go to the Factory web site.*

- *I believe a new type of standard is indicated. This is one that is updated annually and Internet based. Inputs are more like a "blog" that are considered by Industry Experts and then published. Updates follow technical advances, new connectors and, of course, cleaning and inspection products.*

✓ **A strong indicator of "Best Practice" is the ability of the procedure to clean the connection the first time.**

✓ **This is important because existing standards state a connector surface may be cleaned "up to 5 times".**

✓ Most of the folks I have met will change a jumper or pull a circuit card and return it for unnecessary warranty service!

✓ **BEST PRACTICE IS <u>NOT</u> TO "DRY CLEAN FIRST" AND THEN FOLLOW WITH A "WET-TO-DRY".**

POSSIBLE DAMAGE, WASTE OF TIME AND EFFORT, 'DO IT RIGHT THE 1ST TIME'!

✓ **"Refine the Define" !!!**

Questions & Challenges?

Edward J. Forrest, Jr. (ed)
Founder Inventor. Author
RMS (RaceMarketingServices)™
est: 1974
Marietta, Atlanta, Georgia. USA

edforrest@live.com +770-971-8100

www.fiberopticprecisioncleaning.com

Bringing Ideas Together™

www.createspace.com/5371025

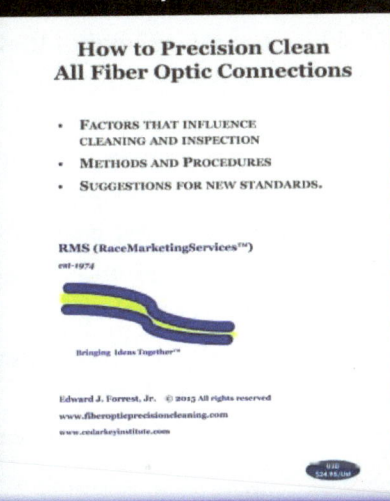

www.createspace.com/5173068

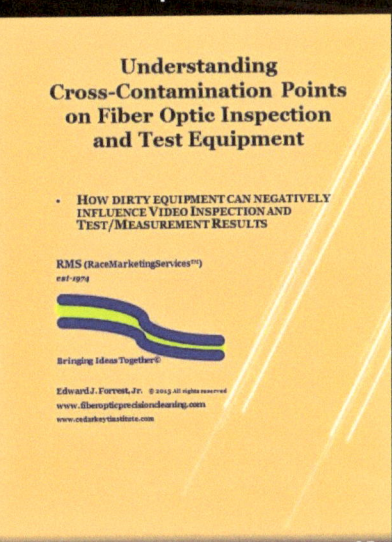

www.createspace.com/5296345

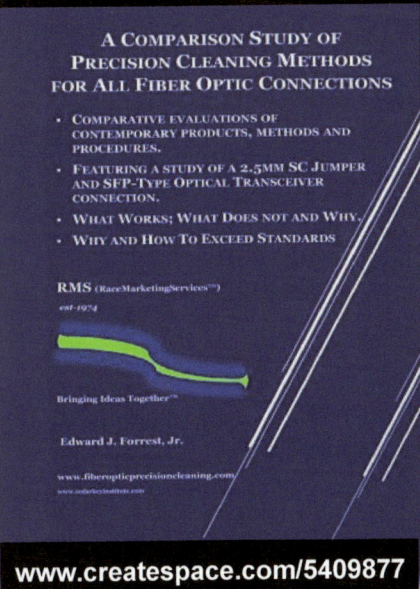

www.createspace.com/5409877

TEST YOUR KNOWLEDGE:

1.) ___T___F   The first thing a technician should do is observe the ambient environment and get an idea of the type of debris or contamination that may be present.

2.) ___T___F: 99.9% isopropyl alcohol (IPA) is the best cleaner because it removes the widest range of debris and contamination all the time.

3.) ___T___F: Cotton and paper are just as good as microfibers and non-woven cellulose/polyester blends to hold and absorb debris and contamination.

4.)___T___F: "Field of View" refers to the area seen by most video inspection

5.)___T___F: Fluidic contamination stays in place and dry debris moves. There is not a transfer between end faces.

6.)___T___F: One problem with swabs is that foam is not as good as other materials to absorb

7.)___T___F: All fiber optic cleaning solvents are the same; they are simply packaged differently

8.)___T___F: If I do not have a video scope, cleaning multiple times works just fine

9.)___T___T: A major advantage of all "probe tools" is their convenience

10.)___T___F: When using a cleaning platform, hold the end face at 90 degree perpendicular to the cleaning surface for best result.

11.) ___T___F: The "best practice" fiber optic cleaning technique is a "Figure-8" motion because it is fast and repetitive.

12.) ___T___F: "Wet to Dry" cleaning does not require video inspection.

13.) ___T___F: The straight line cleaning action moves dry debris and fluidic contamination away from the initial point of contact.

14.) ___T___F: Fiber optic standards, such as IEC 61300-3-35, are updated every 5 years with annual bulletins and updates.

15.) ___T___F: Many of the same cleaning techniques and products typically used for fusion splice prep are appropriate for end face cleaning.

ANSWERS: TEST YOUR KNOWLEDGE:

1.) ___T The first thing a technician should do is observe the ambient environment and get an idea of the type of debris or contamination that may be present.

2.) ___F: 99.9% isopropyl alcohol is not as good a cleaner (solvent) than a precision hydrocarbon, HFE7100 formulas, and new aqueous cleaners.

3.) ___F: Cotton and paper lint, tear and shred. These are good enough to absorb a fluid...but not for precision cleaning. In doubt...don't use them.

4.)___F: "Field of View" is relative and limited by magnification. 400x actually sees less of the end face than 200x. Resolution is an important specific characteristic of a video scope. Higher the magnification, greater the resolution. Look for lower magnification and greater resolution for "best practice".

5.)___F: Fluidic contamination moves and transfers while dry debris stays in place. Dry debris can damage a surface if it has height.

6.)___F: There are literally thousands of types of foam. Be sure you select ones that are "medical" or "cleanroom grade". Foam absorbs and is a great cleaning media.

7.)___F: There are even 15-20 HFE7100 formulas and all of them are not the same cleaning power. Precision hydrocarbons and aqueous cleaners cost about the same as 99.9% IPA and are superior cleaners.

8.)___F: If a video scope is not available...always use the 3rd "hybrid/combination" process as the safety net.

9.)___T: A major advantage of all "probe tools" is their convenience

10.)___T: When using a cleaning platform, hold the end face at 90 degree perpendicular to the cleaning surface for best result.

11.) ___F: The "Figure-8 Motion" is a polishing action. It's not a cleaning action.

12.) ___F: Every connection, no matter which technique, should be video inspected. However, the "hybrid" or "combination" technique works best because the amount of solvent is limited and the surface is dried.

13.) ___T: The detail of this procedure is common sense. The fiber optic surface is so small it is easy to contaminate and   re-contaminate.

14.) ___F: As of 2015, fiber optic standards for precision cleaning and inspection are updated every ten years.

15.) ___F: Fusion splice prep is a completely different application than end face cleaning. For fusion splice, the side and length of the fiber is cleaned. For this, IPA works well enough. Cross-use of IPA to end face cleaning can result in inferior results.

For some, it is difficult to imagine how a topic as mundane as cleaning a fiber optic connection can influence an entire Industry. After all, who has not (and may still) cleaned the surface on a clean t-shirt, or as one technician advised me: "*under the collar of a shirt is better*"!

If you are on a production line, how you are taught to clean the surface not only influences the final product, but also its' installation.   If you are an installer, and you are not able to establish service, you may struggle cleaning the surface four or five times, only to return a component to the producer...only to have a warranty repair embarrassingly rejected because the fiber optic surface was contaminated and not properly cleaned!   For some reason,   within all the brilliance of fiber optic design and deployment,   we seem to have lost sight of an important fundamental: *if the connection is not properly cleaned is will not transmit to design standard and surely will not test with accuracy.* We seem to know more about cleaning a floor than precision cleaning a fiber optic device! Surely, there are as many products!

The Telecommunications Industry is in continual evolution. Not too many years ago cables were laid from the back of horse and buggy: now fiber optic lines are deployed trans-continental, trans-oceanic, and to the home or office desk.   The end user expects reliability and capacity: *what is deployed now, was once considered theoretical only a few years ago.*   New connections are always entering the market: just a few years ago 'expanded beam' was considered 'no clean' until the realization that fluidic contamination can transfer! The same may happen with the humble Jumper-Side SC connection as icing from a morning donut transfers to the backplane and may foul the adapter sleeve during insertion! "Worst case leads to Best Practice".

The results of this study indicate that just about any cleaning product can be made to work!   However, without 100% video-inspection any cleaning procedure is truly a wish and a hope.   It is a hopeful result this study helped others understand these important tenets: 1.) there are many types of potential debris and contamination to consider, 2.) not all cleaning products and associated procedures return the same results, 3.) 'first time cleaning' of some debris or contamination is possible. 4.) "blind cleaning" is far more common than many in the Industry feel comfortable accepting,   5.) precision cleaning each time a connection is 'made' is fundamental best practice, and 6.) video inspection and not (common place) reliance on a light source and power meter is *how to measure cleanliness. Challenge all suppliers of cleaning products by asking for performance comparisons and creating your own internal standards.*

If I can help you with this, please advise your requirements!

All the best, Ed Forrest
September-2015

ABOUT THE AUTHOR:

Thank you for purchasing this book. Ed Forrest has been actively involved in specification and applications engineering of various precision cleaning applications for more than 25 years. Previously employed at ITW Chemtronics®, retired in 2014, he was schooled to analyze precision and gross cleaning applications in a wide range of applications. In 2001 he began development of a program that resulted in formal approvals at all major telecommunications providers.

He has seven patents specifically in the areas or fiber optic precision cleaning with six products in production. He innovated a chemical mid-span break-in for ribbon fiber. He has other patents pending.

Ed is active on fiber optic standards committees and is considered a SME in the study of fiber optic cleaning and inspection. His work is based on field experiences and the needs of designers, crafts persons and production line workers.

His practical thesis of "Five Zone Cleaning" is a look forward to the times when high speed and capacity of fiber optic transmission (even more) will be impacted by a contaminated or improperly cleaned connections. He has uniquely researched inspection of the 4th and 5th Zone and the influences of various debris and contamination positioned on areas of the connector.

As an Electronics Manufacturer's Representative throughout the 1970's, he actively participated in the early introduction of some of the most fundamental electronic products in the changeover from analogue to solid state. These included solid state components, consumer products including hand-held calculators, esoteric high fidelity, test equipment, games and other electronic products (including the 'delightful insanity of CB-radios) considered 'cornerstones' of the contemporary marketplace. He also has production credits in that Industry

Turning to a then-developing market segment in the Home Furnishings Industry, by coordinating North American and International Development, using an effective agency in Denmark he was able to work throughout Europe prior to the time of the EU. Incoordination with C.ITOH (est-1860) , he traveled and developed a Japanese market long before current interest in the important nations of The Pacific Rim. He initiated promotional activity in conjunction with USA Embassies, individual USA states resulting in active trade in Denmark, Sweden, Finland, Italy, Germany. Great Britain, nations in The Middle East and South Africa. Also, he has production credits in that industry segment.

Early career as a Technical Representative in Union Carbide Corporation's Automotive Consumer Products group with career-forming experiences include introduction of Prestone® AntiFreeze as a Summer Coolant in a one year NASCAR race test and associated promotions, as well as, an innovative time with Standard Oil of Ohio® as SOHIO® introduced "self-service fueling" to the market. He competed in this massive market when brands like STP® , BarsLeak®, Simonize® and Wynn's® dominated consumer interest.

He is a hobbyist collector of esoteric high-fidelity, enjoys photography, archaeology, ancient history and raced in SCCA 'wheel to wheel' in more than 300 events in cars of his design...with a little help from his friends! . Married with a fascination for Weimaraners, he and his wife, Lanet, are often on the edge with three lovely specimens. They travel as they can.

- <u>References:</u>

- Residual Contamination on Fiber Optic End Faces from Use of Polar Alcohols Paul Blair and Edward Forrest ITW Chemtronics June 2002 (rev:11/03) White Paper
- HFE-7100/Proprietary Formulation Cleaning Comparison to 99.9% Isopropanol James Fitzgerald. ITW Chemtronics   Research Chemist: 2003. Laboratory Test Soils a.) Animal Fat – Representative of skin and fingerprint oils Multipurpose Grease b.) LUBRIMATIC #11316 Motor Oil c.) Quaker State 10W 40 d.) Silicone Oil Dow Corning representative of pulling lube and buffer gel
- HFE-7100/Proprietary Formulation Cleaning Comparison to Precision Hydrocarbon/Proprietary Formulation James Fitzgerald. ITW Chemtronics Research Chemist: 2003. Laboratory Test Soils a.) Animal Fat – Representative of skin and fingerprint oils b.) Multipurpose Grease – LUBRIMATIC #113163) c.) Motor Oil Quaker State 10W 40 d.) Silicone Oil – Dow Corning representative of pulling lube and buffer gel
- A Study of 99.9% Isopropanol Absorption Rate of Water from Air Susan Max: Lead Chemist. ITW Chemtronics® 2004. Laboratory Test.
- A More Effective Means of Cleaning Fiber Optic Connections in FTTH, Outside Plant, and, OEM Applications. White Paper-Edward J. Forrest FTTH-2005
- Inspection and Cleaning Procedures for Fiber Optic Connections All contents © 1992–2006 Cisco Systems, Inc. Document ID: 51834 8-26-2006
- Soil Removal from End Face Utilizing Cisco Series of Ten Diverse Soils: Paul Blair, Ed Forrest, And Susan Max ITW Chemtronics: 2006. Laboratory Test
- Generating a Static Field When Precision Cleaning a Fiber Optic Connection: Paul Blair, Susan Max, and Edward Forrest. ITW Chemtronics: 2008 Lab Test
- TIA 455-240 September-2009
- IEC 61300-3-35 ed1.0 2009 August 2009
- Telcordia GR-2923-CORE. February-2010
- Contemporary Considerations When Precision   Cleaning Fiber Optic Connections: Performance-Inspection-Environmental Matters: White Paper Edward J. Forrest: November-2010
- SAE AIR 6031. 2012 Cleaning fiber optic connections.
- Comparisons of various cleaning solvents acting on ten complex soils and Investigation of Contamination of the Horizontal and Vertical Ferrule. Laboratory Test recorded on Video. Edward J. Forrest: January 2011
- Interferometer readings courtesy of Promet Corporation using a FiBO® 250 device.
- Contemporary Considerations When Precision Cleaning a Fiber Optic Connection: 2011v8
- Edward J. Forrest: Power Point. "Comparisons of cleaning techniques with audio and video". Training session. 2014
- Bill Woodward "FOI" Fiber Optic Installer (ETA) and published by SYBEX
- "Video Lab Comparisons of Aqueous Cleaning Products and Procedures": Edward J. Forrest, Jr. August-2015
- "VideoLab Tests of Various Cleaning Procedures on simple and complex contaminants." Edward J. Forrest, Jr. August-2015
- "The Need for Processes that Future Proof the Fiber Optic Installation". White Paper. Edward J. Forrest, Jr. July-2015
- "How we do and should not; should and may not, precision clean and inspect a fiber optic connection". Training with video. Edward J. Forrest, Jr. 6/2015
- "An Evaluation of Aqueous Cleaning Processes for Fiber Optic End Face Connections". VideoLab Study of commercial and experimental aqueous cleaners. PowerPoint   Edward J. Forrest, Jr. August-2015

Notes

www.ingramcontent.com/pod-product-compliance
Lightning Source LLC
Chambersburg PA
CBHW040828180526
45159CB00001B/102